D1698513

Himalaya, Dynamics of a Giant 1

*In memory of distinguished professors
M. Gaetani, P. Molnar and A. Steck*

SCIENCES

*Geoscience*, Field Director – Yves Lagabrielle

*Dynamics of the Continental Lithosphere,*
Subject Head – Sylvie Leroy

# Himalaya, Dynamics of a Giant 1

*Geodynamic Setting of
the Himalayan Range*

*Coordinated by*
Rodolphe Cattin
Jean-Luc Epard

WILEY

First published 2023 in Great Britain and the United States by ISTE Ltd and John Wiley & Sons, Inc.

ISTE Ltd
27-37 St George's Road
London SW19 4EU
UK

www.iste.co.uk

John Wiley & Sons, Inc.
111 River Street
Hoboken, NJ 07030
USA

www.wiley.com

Library of Congress Control Number: 2022948342

British Library Cataloguing-in-Publication Data
A CIP record for this book is available from the British Library
ISBN 978-1-78945-129-0

ERC code:
PE10 Earth System Science
 *PE10_5 Geology, tectonics, volcanology*

# Contents

## Chapter 2. Building the Tibetan Plateau During the Collision Between the India and Asia Plates

Anne REPLUMAZ, Cécile LASSERRE, Stéphane GUILLOT, Marie-Luce CHEVALIER, Fabio A. CAPITANIO, Francesca FUNICIELLO, Fanny GOUSSIN and Shiguang WANG

## Chapter 3. The Major Thrust Faults and Shear Zones

Djordje GRUJIC and Isabelle COUTAND

György HETÉNYI, Jérôme VERGNE, Laurent BOLLINGER, Shiba SUBEDI,
Konstantinos MICHAILOS and Dowchu DRUKPA

## Chapter 7. Application of Near-surface Geophysical Methods for Imaging Active Faults in the Himalaya

Dowchu DRUKPA, Stéphanie GAUTIER and Rodolphe CATTIN

## Chapter 8. Overview of Hydrothermal Systems in the Nepal Himalaya

Frédéric GIRAULT, Christian FRANCE-LANORD, Lok Bijaya ADHIKARI, Bishal Nath UPRETI, Kabi Raj PAUDYAL, Ananta Prasad GAJUREL, Pierre AGRINIER, Rémi LOSNO, Sandeep THAPA, Shashi TAMANG, Sudhan Singh MAHAT, Mukunda BHATTARAI, Bharat Prasad KOIRALA, Ratna Mani GUPTA, Kapil MAHARJAN, Nabin Ghising TAMANG, Hélène BOUQUEREL, Jérôme GAILLARDET, Mathieu DELLINGER, François PREVOT, Carine CHADUTEAU, Thomas RIGAUDIER, Nelly ASSAYAG and Frédéric PERRIER

# Tributes

In the past five years, three of our friends, colleagues and mentors Maurizio Gaetani, Peter Molnar and Albrecht Steck have passed away. Their outstanding contribution to the knowledge of the Himalayan range has influenced many authors of this book. We pay tribute to them in the following paragraphs. We dedicate this book to these three exceptional professors.

## The mountains of Asia, and the charm of romantic geology – A tribute to the legacy of Maurizio Gaetani (1940–2017) by Eduardo Garzanti

Student of Ardito Desio, organizer of the 1954 Italian conquest of K2 and younger colleague of Riccardo Assereto, killed by a landslide during the second Friuli earthquake of September 9, 1976, the everlasting love of Maurizio Gaetani for Asian geology began in 1962, with his Thesis fieldwork in the Alborz Mountains of Iran. During summer 1977, as Ladakh opened to foreigners, Maurizio first discovered with Alda Nicora the Cretaceous/Triassic boundary in the Zanskar Range. On August 1, 1981, we took a bus from Delhi to Lahul and crossed with horses and horsemen the Baralacha La and Phirtse La to describe the stratigraphy of the Paleozoic–Eocene succession of the Tethys Himalaya. New expeditions were led by Maurizio to Zanskar in 1984 and 1987 to reconstruct the paleogeographic history of northern India, from the newly identified Early Paleozoic Pan-African orogeny to Upper Paleozoic Neotethyan rifting and the subsequent Mesozoic passive-margin evolution terminated with early Paleogene collision with the Asian arc-trench system.

In the meanwhile, Maurizio's Karakorum adventure had begun with the 1986 expedition to the Hunza Valley. From then on, Maurizio's unique dedication to Karakorum geology is testified by 10 expeditions he led to Chitral, Wakhan, Shimshal and Shaksgam, during which every meter of the stratigraphic section from Ordovician to Cretaceous was measured to reconstruct the opening of Paleotethys and Neotethys and their subsequent closure during early and late Mesozoic orogenies. The amazing amount of work carried out during these surveys is condensed in the magnificent geological map of North Karakorum and summarized in his last paper "Blank on the Geological Map". The legacy of Maurizio Gaetani is to remind us that, no matter how much technology is involved, any scientific adventure is primarily a romantic adventure.

Maurizio Gaetani has contributed to about 50 articles published in peer review journals. Here is a list of his major contributions:

Gaetani, M. (2016). Blank on the geological map. *Rendiconti Lincei*, 27(2), 181–195.

Gaetani, M. and Garzanti, E. (1991). Multicyclic history of the northern India continental margin (northwestern Himalaya). *AAPG Bulletin*, 75(9), 427–1446.

Gaetani, M., Nicora, A., Premoli Silva, I. (1980). Uppermost Cretaceous and Paleocene in the Zanskar range (Ladakh-Himalaya). *Rivista Italiana di Paleontologia e Stratigrafia*, 86(1), 127–166.

Gaetani, M., Garzanti, E., Jadoul, F., Nicora, A., Tintori, A., Pasini, M., Khan, K.S.A. (1990). The north Karakorum side of the Central Asia geopuzzle. *Geological Society of America Bulletin*, 102(1), 54–62.

Zanchi, A. and Gaetani, M. (2011). The geology of the Karakoram range, Pakistan: The new 1: 100,000 geological map of Central-Western Karakoram. *Italian Journal of Geosciences*, 130(2), 161–262.

## Always on the cutting edge and looking for new ideas to advance Earth Sciences – Peter Molnar (1943–2022) by Vincent Godard, Rodolphe Cattin and György Hetényi

In the final phase of preparing the three volumes of this book, we sadly learned of the passing of Peter Molnar. Peter was a giant of the Earth Sciences, whose contributions would be too long to list exhaustively in this tribute; he left a remarkably enduring mark on Himalaya–Tibet research.

Peter's scientific career began at a pivotal moment for Earth Sciences. He has significantly contributed to developing and applying the new paradigm of Plate Tectonics. Using innovative approaches based on tectonics, seismology, paleomagnetism and satellite imagery, Peter revolutionized our understanding of continental deformation and lithosphere behavior. Although Peter worked on a wide range of problems and geoscientific contexts, the India–Asia collision and the dynamics of the Himalaya–Tibet system have very often been at the core of his investigations. He has left a lasting imprint on research in this region, particularly by his desire to understand the mechanical processes at work during the deformation of this orogenic system.

Consistent with his comprehensive and integrated approach to geodynamic problems, Peter explored a wide range of ideas and processes related to the Himalaya–Tibet system's global role, and initiated several ideas and research directions that are still active today. One example, among many, is the study of the physical relationships and interaction of mechanisms between the development of Tibet's topography and the Southeast Asian monsoon regime. Among the research fields initiated by Peter, understanding the relationships between erosion, tectonics and climate is undoubtedly one of the most innovative and impactful for our community. Following a series of seminal articles by Peter and his colleagues, the complex interactions between the processes responsible for topographic relief creation and destruction are still actively debated in the Himalaya. Peter's research always focused on a global understanding of these processes, particularly those responsible for the variations in erosion and global sedimentary fluxes related to the evolution of Himalayan orogeny and the late Cenozoic evolution of climate. The fact that so many of these topics are still at the forefront of current research by so many groups worldwide is a major testimony to Peter Molnar's prescience on the dynamics of the Himalaya–Tibet system.

Beyond these outstanding contributions, Peter will be remembered for his great sense of humor and for having always been accessible and available to discuss new ideas with young scientists.

Peter Molnar has contributed to an impressive number of publications, some of which have been milestones in the understanding of the Tibet–Himalaya system:

Bilham, R., Gaur, V.K., Molnar, P. (2001). Himalayan seismic hazard. *Science*, 293(5534), 1442–1444.

Gan, W., Molnar, P., Zhang, P., Xiao, G., Liang, S., Zhang, K., Li, Z., Xu, K., Zhang, L. (2022). Initiation of clockwise rotation and eastward transport of southeastern Tibet inferred from deflected fault traces and GPS observations. *Geological Society of America Bulletin*, 134(5–6), 1129–1142.

Houseman, G.A., McKenzie, D.P., Molnar, P. (1981). Convective instability of a thickened boundary layer and its relevance for the thermal evolution of continental convergent belts. *Journal of Geophysical Research: Solid Earth*, 86(B7), 6115–6132.

Molnar, P. (2012). Isostasy can't be ignored. *Nature Geoscience*, 5(2), 83–83.

Molnar, P. and England, P. (1990). Late Cenozoic uplift of mountain ranges and global climate change: Chicken or egg? *Nature*, 346(6279), 29–34.

Molnar, P. and Tapponnier, P. (1975). Cenozoic Tectonics of Asia: Effects of a continental collision: Features of recent continental tectonics in Asia can be interpreted as results of the India-Eurasia collision. *Science*, 189(4201), 419–426.

Zhang, P.Z., Shen, Z., Wang, M., Gan, W., Bürgmann, R., Molnar, P., Wang, Q., Niu, Z., Sun, J., Wu, J. et al. (2004). Continuous deformation of the Tibetan Plateau from global positioning system data. *Geology*, 32(9), 809–812.

## Geology of the Indian Himalaya – Albrecht Steck (1935–2021) by Jean-Luc Epard and Martin Robyr

Albrecht Steck was the main driving force behind the Himalayan geological research program led at the University of Lausanne over the last 40 years. His work has focused on the Indian Himalaya, particularly on the Mandi to Leh transect. He has directed or supervised eight doctoral theses distributed along this transect. One of the mottos of Albrecht was that any good geological work always starts with a sound geological mapping. Whether in the Alps or in the Himalaya, Albrecht's work excels by the quality of his geological maps. It is becoming a hallmark of Albrecht's work. Indeed, the numerous field missions he led or supervised in Lahul, Zanskar and Ladakh regions allowed the achievement of detailed geological maps covering a remarkable large area of the NW Indian Himalaya.

The research of Albrecht Steck is characterized by the combination of field observations (mapping, stratigraphy, structures and metamorphism) in order to decipher the geometry, kinematics and tectono-metamorphic history associated with orogenic processes. His scientific approach combining a variety of field, structural, petrographic and analytical methods is a hallmark of Albrecht's research. Two publications reflect particularly well the research works led by Albrecht Steck. The first (Steck et al. 1993) concerns a transect from the High Himalayan Crystalline of Lahul to the south to the Indus suture zone to the north; the second (Steck et al. 1998) focused on a complete geological transect through the Tethys Himalaya and the Tso Morari area. The results of these expeditions are also synthesized in a general publication of the Geology of NW Himalaya (Steck 2003). For Albrecht Steck, geology must be made in the field and out of the main touristic roads. His long-term commitment to the detailed study of the NW part of the Himalaya of India significantly contributed to a better understanding of the geology of this region. With Albrecht Steck, the Alpine and Himalayan geological community has lost an eminent researcher and a true Nature lover.

Albrecht Steck has contributed to many geological maps and articles published in peer-reviewed journals. Here are three of his major contributions:

Steck, A. (2003). Geology of the NW Indian Himalaya. *Eclogae Geologicae Helvetiae*, 96, 147–196.

Steck, A., Spring. L., Vannay, J.-C., Masson, H., Stutz, E., Bucher, H., Marchant, R., Tièche, J.-C. (1993). Geological transect across the Northwestern Himalaya in eastern Ladakh and Lahul (A model for the continental collision of India and Asia). *Eclogae Geologicae Helvetiae*, 86(1), 219–263.

Steck, A., Epard, J.-L., Vannay, J.-C., Hunziker, J., Girard, M., Morard, A., Robyr, M. (1998). Geological transect across the Tso Morari and Spiti areas: The nappe structures of the Tethys Himalaya. *Eclogae Geologicae Helvetiae*, 91, 103–121.

# Foreword

**Rodolphe Cattin[1] and Jean-Luc Epard[2]**

*[1] University of Montpellier, France*
*[2] University of Lausanne, Switzerland*

The Himalaya is well known as the largest and highest mountain belt on Earth, stretching 2,500 km from the Nanga Parbat syntaxis in the northwest to the Namche Barwa syntaxis in the southeast, with peaks exceeding 8,000 m in altitude. Resulted from the ongoing collision between the India and Asia plates, the Himalaya is frequently used as the type example of a largely cylindrical mountain belt with a remarkable lateral continuity of major faults and tectonic units across strike.

Advances in geoscience over the past few decades have revealed a more complex picture for the dynamic of this giant, with open questions about the initial stages of the Himalayan building, lateral variations in its structures, variations in tectonic forcing, tectonic–climate coupling and assessment of the natural hazards affecting this area.

In this book, we present the current knowledge on the building and present-day behavior of the Himalayan range. The objective is not to be exhaustive, but to give some key elements to better understand the dynamics of this orogenic wedge. The three volumes of this book present (1) the geodynamic framework of the Himalayan range, (2) its main tectonic units and (3) its current activity. The chapters and volumes in this book are self-contained and can be read in any order. However, the three volumes are linked and provide together a self-consistent image of the Himalayan dynamic at various temporal and spatial scales.

Volume 1, entitled *Geodynamic Setting of the Himalayan Range*, addresses the tectonic framework of the Himalaya and Tibet, the segmented nature of the Himalayan belt and gives two examples of studies focused on near-surface active fault imagery in Bhutan, and hydrothermal system in Nepal.

This volume is coordinated by Rodolphe Cattin (University of Montpellier, France) and Jean-Luc Epard (University of Lausanne, Switzerland) with the help of the editorial team composed of Laurent Bollinger (French Alternative Energies and Atomic Energy Commission, France), György Hetényi (University of Lausanne, Switzerland), Vincent Godard (University of Aix-Marseille, France), Martin Robyr (University of Lausanne, Switzerland) and Julia de Sigoyer (University of Grenoble, France).

All royalties allocated to the authors of this book will be donated to the "Seismology at School" program (see the next pages).

# Preface

# From Research to Education: The Example of the Seismology at School in Nepal Program

György HETÉNYI[1] and Shiba SUBEDI[1,2]

*[1] Institute of Earth Sciences, University of Lausanne, Switzerland*
*[2] Seismology at School in Nepal, Pokhara, Nepal*

Scientific research aims at observing and understanding processes, and enriching our knowledge. But who is in charge of transferring this knowledge to society, to everyday life? Can we expect any researcher to become a company CEO, an engineer, a policy maker or a teacher? Our answer is no, not necessarily, but efforts can be made in that direction, and there are successful examples.

In the context of Himalayan geoscience research, a tremendous amount of information exists. It cannot be all simplified and all translated to local languages of the Himalaya; nevertheless, we found it essential that such knowledge transfer starts. In the aftermath of the 2015 magnitude 7.8 Gorkha earthquake, through a series of fortunate steps, we found ourselves putting down one of the bricks of knowledge transfer by initiating the Seismology at School in Nepal program. Our primary pathway choice was education: in the short-term, raising earthquake awareness and better preparedness can spread through the students to their families, relatives and acquaintanceship; in the longer-term, it is today's school students who will build the next generation of infrastructures.

The program started following a bottom-up approach, with direct cooperation with local schools in Nepal. This ensured motivated participants and direct feedback on the activities and about the needs. The program stands on two main pillars and is described in detail in Subedi et al. (2020a). First, earthquake-related topics have been synthesized and translated to Nepali, together with a series of hands-on experiments, and the local teachers have been trained so that they can teach these in their classes. Second, we have installed relatively cheap seismometers (RaspberryShake 1D) in local schools, which became part of the classroom activities and also recorded waves from earthquakes. This has sparked interest in schools, and the openly and publicly available waveform data is useful for monitoring and research as well. To more closely link these two, we have written a simple earthquake location tutorial that is feasible with typical school computers in Nepal (Subedi et al. 2021).

The program has started in Nepal in 2018; as of 2019, more than 20 schools and seismometers have been involved in the program, and the number is reaching 40 in 2022. There is measurable improvement in students' knowledge (Subedi et al. 2020b), and the feedbacks are very positive. Parallel to classical educational pathways, a series of other activities have been developed in the Seismology at School in Nepal program. Each school has received an Emergency Meeting Point sign in Nepali language. Over 6,000 stickers reminding about earthquakes have been distributed to increase awareness (see Figure 8.10 in Volume 3 – Chapter 8). An Earthquake Awareness Song has been written and composed, and became popular on YouTube (https://www.youtube.com/watch?v=ymE-lrAK0TI). We studied the Hindu religious representation and traditional beliefs about earthquakes (Subedi and Hetényi 2021). Recently, we have developed an educational card game to improve the practical preparation and reaction to earthquakes. Finally, we maintain a website with all information openly available (http://www.seismoschoolnp.org).

The program has so far run on funding that is considered to be small in the research domain, and this has covered the cost of materials and the work in Nepal. More recently, a crowd-funding campaign has been started and evolved successfully – we are very grateful to all funders and donators! In the future, the program aims at growing further, all across Nepal and hopefully all along the Himalaya. This will require more manpower and more funds. The authors of this book have generously given their consent to transfer all royalties to the Seismology at School in Nepal program thank you very much!

There is a strong similarity between this book and the Seismology at School in Nepal program: they both aim at taking research results and carrying them to non-specialists. This book is planned to be published in several languages, and to reach students and interested people around the world. The educational program in Nepal aims at bringing earthquake knowledge to those who really need it as they live in a high hazard area. Both efforts aim at increasing awareness, and, thereby, we hope and wish that their effects reach further across all society.

October 2022

## References

Subedi, S. and Hetényi, G. (2021). The representation of earthquakes in Hindu religion: A literature review to improve educational communications in Nepal. *Front Commun*, 6, 668086. doi:10.3389/fcomm.2021.668086.

Subedi, S., Hetényi, G., Denton, P., Sauron, A. (2020a). Seismology at school in Nepal: A program for educational and citizen seismology through a low-cost seismic network. *Front Earth Sci*, 8, 73. doi:10.3389/feart.2020.00073.

Subedi, S., Hetényi, G., Shackleton, R. (2020b). Impact of an educational program on earthquake awareness and preparedness in Nepal. *Geosci Commun*, 3, 279–290. doi:10.5194/gc-3-279-2020.

Subedi, S., Denton, P., Michailos, K., Hetényi, G. (2021). Making seismology accessible to the public in Nepal: An earthquake location tutorial for education purposes. *Bull Nep Geol Soc*, 38, 149–162.

# PART 1

# Tectonic Framework of the Himalaya and Tibet

# 1

# Plate Reconstructions and Mantle Dynamics Leading to the India–Asia Collision

**Gweltaz MAHÉO[1] and Guillaume DUPONT–NIVET[2]**

*[1] Laboratoire de Géologie : Terre, Planètes et Environnement,
Claude Bernard University, Lyon, France*

*[2] Géosciences Rennes, University of Rennes, France*

## 1.1. Introduction

The origin of the Himalayan range and Tibetan Plateau, the highest morphological feature of the Earth, has always been highly debated. Following the early thermal contraction models of the 19th century, Argand (1924) was the first to suggest that the Himalaya and Tibetan Plateau orogenesis resulted from the convergence of the Indian and Eurasian continents: *le jeu dominant, c'est le rapprochement de deux serres continentales, l'Indo-Afrique et la vieille Eurasie, avec rétrécissement de la Téthys*[1]. In this early view, India (with Africa) and Eurasia continents (named Sial) are moving on a "Sima" basement (previous name of the mantle) and separated on oceanic basin the Tethys (Figure 1.1). This view was applied by Wegener (1929) in his famous

---

1. The dominant game is the approach of two continents, Indo-Africa and old Eurasia, with a shrinking of the Tethys.

*Himalaya, Dynamics of a Giant 1*,
coordinated by Rodolphe CATTIN and Jean-Luc EPARD.
© ISTE Ltd 2023.

continental drift model. Wegener also recognized the long journey of India across the Tethys Ocean. Based on evidence of Carboniferous glaciation on the Indian continent, and the lack of such features in Asia, he assumed that India used to be located close to the South Pole while Asia was in the northern hemisphere. Most ensuing models then related the Himalayan building with the shortening of the Tethyan "geosyncline" during the northward migration of the Indian continent across the Tethys Ocean, and recognized the remains of these oceanic domains in the rocks constituting the "Tethyan Himalaya Unit"(Gansser 1964; Holmes 1965, see Volume 2 – Chapter 3)

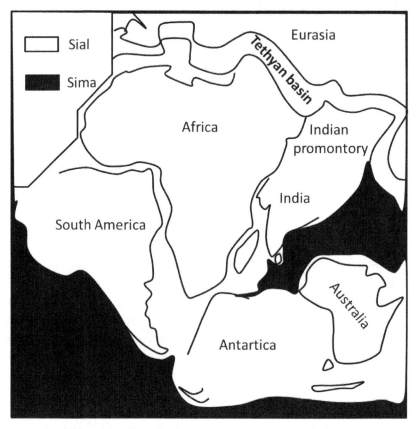

**Figure 1.1.** *Gondwana configuration before the Atlantic Ocean opening. Redrawn after Argand (1924)*

Then came the plate tectonic model revolution conceptualized by Holmes (1965), initiated by Hess (1962), observations of marine magnetic anomalies and their interpretation to infer tectonic plate motion by Vine and Matthews (1963). Dewey and Bird (1970) proposed that the Tethyan basin was actually made of oceanic crust, recognized oceanic rocks in the Himalaya as remnants of oceanic crust (later called ophiolites) trapped in suture zones between colliding continental plates following closure of the "Neotethys Ocean" (see Volume 2 – Chapter 2).

The first paleogeographic reconstruction of India's northward migration across the Neotethys includes early paleomagnetic results and the development of the global tectonic plate circuits based on marine magnetic anomalies (see section 1.2.1 for methodology; Pozzi et al. 1982; Patriat and Achache 1984; Molnar and Tapponnier 1995). These reconstructions allowed estimating throughout geological history, the rate and direction of the Indian lithosphere displacement with respect to Eurasia (Figure 1.2). A slowdown of India–Asia convergent around 50 Ma was readily interpreted as the timing of continental collision (Figure 1.2; Patriat and Achache 1984).

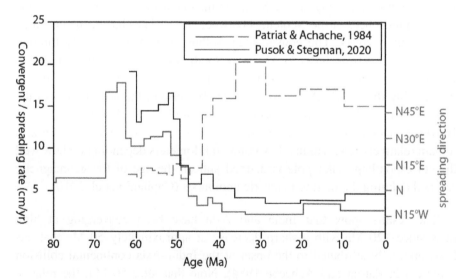

**Figure 1.2.** *Examples of the spreading rate and direction of the Central Indian Ridge (India–Africa) reconstructions based on marine magnetic anomalies data. For a color version of this figure, see www.iste.co.uk/cattin/himalaya1.zip*

Since this pioneering work, the development of geophysical techniques and the accumulation of geological data has led to refinements in kinematic plate reconstructions and continental deformation allowing for precise models of the paleogeographic evolution and involved geodynamic processes.

In this chapter, we will review current data and models proposed for the India–Asia convergence history, including the age of the continental collision and the geodynamic configuration.

## 1.2. The India–Asia convergence and the age of the collision

### 1.2.1. *The India–Asia convergence*

The relative motions of tectonic plates are estimated through time using plate kinematic reconstructions built on the basis of "Euler poles" determined from "marine magnetic anomalies" globally defining a "plate circuit" (Hellinger 1981; Cox and Hart 1986; Torsvik et al. 2012; Müller et al. 2016). Interestingly, in the case of India and Eurasia since 120 Ma, the relative motion cannot be directly measured by marine magnetic anomalies because the oceanic lithosphere of the Neotethys between these plates has been consumed at subduction zones. The convergence is thus estimated indirectly through the plate circuit from India to Africa to South America to North America to Eurasia. Uncertainties thus accumulate and, for example, there are issues in determining the Indian–African kinematics because of the non-rigidity of the eastern African plate separating with the Nubian plate (e.g. Royer et al. 2006). However, uncertainties on the relative positions between India and Eurasia remain relatively low, within a few hundred kilometers depending on how well the corresponding Euler pole is defined and on the age of the geomagnetic reversal defining the marine magnetic anomalies (Capitanio et al. 2010).

The results show that India and Asia have been converging at high rates since 120 Ma with a sharp decrease at approximately 50 Ma that has been originally attributed to the onset of the India–Asia continental collision (Figure 1.2; Patriat and Achache 1984). Note that after 50 Ma, the relative convergence until today amounts to ca. 4,000 km. This implies that, if 50 Ma is indeed the age of the continental collision, this amount of the Indian and Asian continental lithosphere has disappeared in processes such as compressive deformation, continental subduction and lithospheric extrusion and detachment as will be discussed further below. Therefore, kinematic models are constantly being improved to quantify in detail the India–Asia convergence in order to further infer geodynamic processes involved in the continental collision (e.g. Copley et al. 2011; Capitanio et al. 2015; Webb et al. 2017; Schellart et al. 2019). Notably, sharp variations in the convergence have been interpreted to relate to more complex collision models discussed further below (Van Hinsbergen et al. 2012; Jagoutz et al. 2015; Pusok and Stegman 2020, see section 1.4.3 for discussion)). Also, convergence variations have been related to deep processes such as plumes (van Hinsbergen et al. 2011a) or slab break-off (Mahéo et al. 2009), to the nature of ocean floor sediments facilitating subduction (Behr and Becker 2018) or to thickening of the Asian lithosphere hindering subduction (Molnar and Stock 2009). In addition, changes in the convergence direction implying a rotation of India with respect to Asia (Figure 1.2) have been interpreted as resulting from the diachronous collision starting in the West and finishing in the East (e.g. Klootwijk et al. 1992; Patzelt et al. 1996; Hu et al. 2016). Other potential origins include progressive West to East tearing of the subducting Neotethys slab, breaking off the India plate following the continental plate collision (Replumaz et al. 2010) or to surface processes such as monsoons along the Himalayan front (Iaffaldano et al. 2011; Husson et al. 2014).

### 1.2.2. *The age of the India–Asia collision*

When India collided with Asia has been highly debated in the last decade and remains a major controversy. First of all, the collision age must be defined because the collision involves many processes that may be taking place at different times. The collision age considered here is the time at which the converging continental plates come into contact. This initial contact is potentially associated with the cessation of marine deposition, ophiolite obduction, continental subduction and crustal deformation of the underthrusting plate. The processes leave traces in the rocks that can help decipher their precise chronology of what happened during the collision. In the Indo–Asia convergent zone, the collision has been originally assumed to have occurred along the Indus–Yarlung suture zone (Allègre et al. 1984). This suture zone is characterized by remnants of oceanic lithosphere between the India and the Asian plates. It now separates the Lhasa terrane (the southernmost terrane of Asia, or "Greater Asia") from the Tethyan Himalaya sediments that are generally interpreted to represent the northern passive margin of India, or "Greater India" (e.g. Figure 1.3).

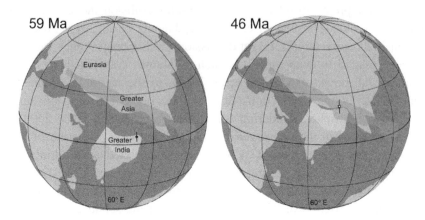

**Figure 1.3.** *India–Asia collision. The yellow parts of the continents are precisely constrained with the global plate circuit based on marine magnetic anomalies (see the text). The past continental extents are constrained by paleolatitude estimates from paleomagnetic data recovered from rocks of the "Greater India" and "Greater Asia" fragments (black and white dots respectively; modified from Dupont-Nivet et al. 2010). The gray shading indicates a continuous seismic tomography positive anomaly in the mantle, interpreted as the remnant of the subducted lithospheric slab of the Neotethys, that would have broken off at the time and location of the collision (see the text). For a color version of this figure, see www.iste.co.uk/cattin/himalaya1.zip*

Various methods are used to constrain the timing of India–Asia contact (see DeCelles et al. (2014), Hu et al. (2016), Kapp and DeCelles (2019) for detailed reviews).

## 1.2.2.1. *Decrease in India–Asia convergence rates*

As previously discussed, reconstructions of the Indian plate kinematics based on paleomagnetic data indicate significant variation of the convergent rates with Asia. The evidence of a sharp decrease in the convergent rate at about 55–50 Ma was originally attributed to the collision with the onset of subduction of the Indian Continental margin beneath the Asia plate slowing down the convergence (Figure 1.2; Patriat and Achache 1984). However, deceleration of convergent rates may also be related to other features such as interaction with the Deccan Plume (van Hinsbergen et al. 2011a) or breakoff of the Neotethys oceanic lithosphere which was driving the continental subduction (Mahéo et al. 2009) as discussed above.

### 1.2.2.2. *Reconstruction of relative plate positions*

Another way to estimate the timing of collision based on paleomagnetism is to reconstruct in time the past latitudinal positions of rocks preserved in the collision zone that formed on the northern margin of Indian and on the southern margin of Asia before and during the collision (Figure 1.3; see the paleomagnetic methods in detail in 1.3.1). The age at which both paleolatitude start to overlap is thus interpreted as the age and paleolatitude of collision. This powerful approach enables us to position past continental slivers involved in the collision, but by itself it cannot resolve if these slivers were really at the end of Indian and Asian promontories or if they were separated from these continents by large oceans.

### 1.2.2.3. *Provenance of detrital sediments*

As the continent margins of India and Asia are coming into contact, detrital sediments may travel across the future suture zone. The provenance of detrital material usually bears characteristic features that enable us to distinguish its source continent, based on mineralogy, coupled with mineral (especially zircon) geochemical composition and ages (e.g. Critelli and Garzanti 1994; DeCelles et al. 1998). The most recent estimates based on detrital evidence of Asian detritus onto the India passive margin as it arrived in the collision zone pinpoint the collision age precisely ca. 58–60 Ma (e.g. DeCelles et al. 2014; Hu et al. 2016; Najman et al. 2017; Garzanti et al. 2018; Garzanti 2019). However, although the age is well constrained, the source of some detrital sediments is highly debated because its Asian provenance, characterized by distinct signatures from the Yarlung Zangpo suture, may be ambiguous. Indeed, several authors suggest that some of these distinctively Asian sediments may actually be sourced from intraoceanic units, such as the Kohistan–Ladakh arc. The latter came into collision first with India coming northward at about 55–60 Ma and only later with Asia at about 40–50 Ma (Bouilhol et al. 2013; Jagoutz et al. 2015; Bhattacharya et al. 2020; Martin et al. 2020). Such interpretation has strong incidences on plate reconstructions that will be discussed in the following section.

### 1.2.2.4. *Paleontological records*

Records use Indian fossil first occurrences in Asia to evidence potential land connections for biotic interchange between Asia and India (e.g. Jaeger et al. 1989; Ali and Aitchison 2008). However, changes in landmass connection related to the sea-level change as well as plate tectonic may have induced more complex biological interactions than previously thought (Chatterjee and Scotese 2010). More recently, these results have been combined with statistical analyses of molecular phylogenies showing that interchange can significantly pre- or post-date the collision time depending on various taxon transports and the occurrence of inter-continental weepstake dispersals through rafting (e.g. Krause et al. 2019). In fine, the estimates are relatively imprecise and more relevant to document the impact of the collision on biotic evolution than the opposite.

### 1.2.2.5. *Continental subduction*

Following collision, the Indian north continental margin is temporally subducted beneath India as a consequence of the subducting Neotethys ocean slab drag, taking rocks to great depths where minerals are formed and transformed at so-called ultrahigh pressures (UHP). Some of these rocks are later exhumed and can be collected today in the collision zone. Then, the timing of the UHP Indian continental rock formation can be dated with the associated thermochronological systems and these ages will provide an upper bond for the onset of collision. Such rocks are only preserved in western Himalaya, as other high-pressure rocks from central Himalaya are relicts partially retromorphosed in granulite facies conditions during collision that has obliterated most of the UHP minerals (see Guillot et al. (2008), and Volume 2 – Chapter 4 for details). The ages of the UHP metamorphism range from 53 to 46 Ma, thus imply that continental subduction of the north Indian margin was already ongoing at that time in western Himalaya and thus that continental collision had already occurred. Note that this timing is consistent with the slowdown of the India convergence rate that was indeed attributed to continental subduction. However, the question remains if this metamorphism is recording the subduction of India beneath Asia or beneath some intraoceanic arc.

### 1.2.2.6. *Magmatism (cessation and onset)*

Magmatic pulses and associated sources generated by the subduction of various materials during the convergence and collision may also help reconstruct the geodynamic evolution and in turn constrain the timing of collision. The closure of the Neotethys Ocean by subduction is related to several magmatic episodes, especially the emplacement of the Gangdese or Transhimalaya arc along the south Asian active continental margin (Allègre et al. 1984; Kapp and DeCelles 2019). However, the magmatic evolution of the Gandese arc is relatively complex with arc-like magmatic rocks emplaced from Paleocene to Miocene time (e.g. Chung et al. (2005), Zhu et al. (2015), Ma et al. (2022), for review). Actually, several processes may explain this magmatism including oceanic and continental subduction, slab breakoff, roll-back or delamination. Using the magmatism to constrain the transition from oceanic to continental collision is thus complex and led to various collision estimates ranging from 40 Ma to 60 Ma (see Hu et al. (2016), for detailed review).

### 1.2.2.7. *Sedimentary evolution*

Contact between previously separated continents and ongoing crustal deformation will have significant impact on sedimentary evolution and basin dynamics. Sediment records related to collision include cessation of oceanic sedimentation, change in basin dynamic recorded by major unconformities as well as the developement of a foreland basin on the India continent or detrital sedimentation on the suture zone. These sedimentary records may postdate collision, but some, such as cessation of marine sedimentation, may occur several million years after collision depending on sea level changes. One of the most convincing and well-constrained sedimentary records of the collision is the developement and evolution of a foreland basin related to Indian plate flexuration following continental subduction as well as ophiolite obduction (Garzanti et al. 1987; Beck et al. 1996; DeCelles et al. 2002). Detailed stratigraphic studies in the western Himalaya (e.g. Garzanti et al. 1987; Guillot et al. 2003) evidence sedimentation change at about 55 Ma related to flexuration of the Indian Continental margin beneath the obduction Neotethys ophiolite. This implies that Indian continental subduction was already ongoing at about 55 Ma as suggested by UHP rock age as well as convergent rate decrease and thus place collision before 55 Ma. Similarly, stratigraphic data from the westernmost Himalaya (Beck et al. 1996; Rowley 1996) evidence that collision occurred as early as late Cretaceous, leading to the suggestion of a West to East diachronous collision, as confirmed by the more recent stratigraphic compilations along the Himalaya (Hu et al. 2016).

### 1.2.2.8. *Crustal deformation*

During the collision, the Indian continental crust underthrusted beneath the Asia margin (or beneath an intraoceanic arc) records significant deformation that can be quantified and dated. Especially south verging deformation (top to the south thrusting and recumbent folding) within the Indian continental margin sedimentary cover, the Tehyan Himalaya is expected. The oldest of such shortening-related structures and contemporaneous metamorphism are comprised between 44 and 54 Ma (seeHu et al. (2016), for review). 66 to 55 Ma accretionnay-wegde structures have also been recognized in the Tethyan Himalaya sediments from northwestern Pakistan (Beck et al. 1995). Once again, potential arc-continent collision may also account for such features.

In summary, several methods suggest that India–Asia collision occurred at about 60–55 Ma. However, earliest (65 Ma) and latest (40 Ma) timings were also proposed. The convergence between India and Asia has been originally mostly viewed as a simple convergence associated with a unique subduction zone. However, as already suggested above, more complex settings have been proposed in approximately the last 15 years. Notably the Aitchison et al. (2007) seminal study involving possible intraoceanic subduction zones and a related intraoceanic arc that may have come into collision with the Indian continent as it was arriving from the south. In that view, India would then continue to move pushing northward the intraoceanic arc at its front until reaching the southern margin of the Asian continent. These ideas have led to a resurgence of models, data and controversies to better reconstruct the India–Asia collision. Crucial to testing these various models are the aforementioned methods to reconstruct the chronology of events associated with the collision and the position and shape of the Indian and Asian continents before collision. We now review the plate configuration models in the following two sections and discuss their implications on the current debate on the evolution of the India–Asia collision.

## TO GO FURTHER ON PALEOMAGNETISM.

The tectonic plate circuit, based on retro-fitting marine magnetic anomalies of the same age, enables us to constrain only provides the motion of large tectonic plates relative to each other. However, the position through time of the plate circuit with respect to the Earth's axis still needs to be defined. On geologic time scales, the geographic north it is indistinguishable from the position of the Earth's magnetic pole which can be determined with paleomagnetic data properly acquired (Butler 1992). Because of tectonic plates moving, the resulting paleomagnetic poles appear to move through geologic time on a path called the apparent polar wander path (APWP). Using the plate circuit, the paleomagnetic data from all tectonic plates can be rotated into a synthetic global APWP (Besse and Courtillot 2002; Torsvik et al. 2012). This enables us to combine paleomagnetic datasets from all plates and reduce uncertainties on the position of the plate circuit with respect to the Earth's axis. An alternative method consists of using the position of plates with respect to hotspot tracks (e.g. Müller et al. 2018). But hotspots in the mantle move with respect to the Earth's axis and still require to be referenced using paleomagnetic data from hotspots.

The plate circuit and associated global APWP are particularly important to place quantitative constraints on amounts and rates of shortening accommodated in the Himalaya–Tibetan orogen, as well as to identify ages of the India–Asia collision. These arguments eventually lie at the basis of spectacular geodynamic and tectonic processes, such as tectonic extrusion, the possibility of many hundreds of kilometers of continental subduction, and consumption of well over 1,000 km of a continental overriding plate during ocean closure and continent–continent collision. However, the synthetic APWP remains based on too limited low amounts of data. As a result, sliding windows of 10, 20 or even 50 Ma are used, filtering out "details" in plate motions. Recent efforts have identified issues in the statistical methods estimating uncertainties on the APWP and are actively working on developing new methods incorporating various datasets to reduce APWP uncertainties and ultimately better define the India–Asia convergence evolution (Rowley 2019; van Hinsbergen et al. 2021; Vaes et al. 2022).

## 1.3. Plate collision configurations

### 1.3.1. *Reconstructing lost continental margins*

During the convergence and the collision, portions of the north Indian and south Asian continental margins have been lost mainly through continental subduction (India) or crustal shortening. These processes must be taken into account to restore the initial shape of the colliding continents and thus better estimate the timing of collision. This can be done with various methods, in particular structural geology to estimate the amount of crustal shortening and paleomagnetism to reconstruct past latitudes of continental blocks involved in the collision.

Early view of the pre-Himalayan India by Argand (1924) already suggested that a vast continental domain, the Indian promontory, disappeared under Asia during collision (Figure 1.1). With the later recognition of the existence of the Neotethys Ocean subduction, the size of such a promontory was significantly reduced. However, the initial size of a Greater India and Greater Asia that disappeared after the collision is still highly debated because of important implications on the geodynamic models that best describe the collision (e.g. Aitchison et al. 2007; Jagoutz et al. 2015). As a result, considerable efforts have been made in constraining the past configuration of the collision zone, usually performed by combining datasets from structural geology and paleomagnetism (e.g. Chen et al. 1993; Van Hinsbergen et al. 2011b).

The vast amount of structural data available from the various components of the collision zone, mostly shortening and strike-slip, has been compiled to provide quantitative estimates of how much convergence may be accommodated by crustal deformation and where (e.g. Dewey et al. 1988; Guillot et al. 2003; Van Hinsbergen et al. 2011b). These generally constitute minimum estimates as cumulative structural offset data is rarely exhaustive and usually fails to integrate ductile deformation. Nevertheless, van Hinsbergen et al. (2011a), end up with 600–1,000 km of cumulated shortening for Greater Asia and about 1,000 km for Greater India since the collision. These values are particularly interesting when compared to the convergence estimates from the plate circuits (Huang et al. 2015; van Hinsbergen et al. 2019). These values virtually exclude simple continental collision models since 50 Ma, let alone 60 Ma, imply more than the double convergence to be accommodated by continental crust deformation. Further insight is gained by comparing paleomagnetic data from Greater India and Greater Asia.

For small fragments of tectonic plates (terranes) that are not incorporated in the plate circuit (such as Greater India and Greater Asia), paleomagnetism enables us to estimate their paleolatitude and orientation with respect to North, by comparing the observed magnetic directions recorded in rocks when they formed. In practice, thousands of samples are collected to obtain an average paleomagnetic direction recorded by rocks from Greater India and from Greater Asia near the collision age (Dupont-Nivet et al. 2010; Yuan et al. 2021, e.g.). The average inclination of these directions enables us to estimate the original latitude of these rocks because the Earth's magnetic field behaves as a geocentric dipole generating a field with increasing inclination from the equator to the poles (Butler 1992; Tauxe 2010). The obtained latitudes of Greater Indian and Greater Asian rocks can then be compared to the latitude expected for the India and Asian continents, respectively, which are constrained by the global paleomagnetic database combined with the tectonic plate circuit based on fitting marine magnetic anomalies (see section "TO GO FURTHER WITH PALEOMAGNETISM"). The difference in latitudes, with associated uncertainties, is used to estimate the extent of Greater India and Greater Asia (Figure 1.3). Similarly, the paleomagnetic declination recording the past north direction in the rocks when they formed can be used to estimate by how much these rocks rotated with respect to Asia or India. These rotations can help estimate the deformation of these terranes and identify tectonic processes associated with them such as strike slip deformation along large faults. Despite the large amount of paleomagnetic data, uncertainties remain of the order of $\pm 5°$ (ca. $\pm 500$ km in latitude) due to the inherent noise in the magnetic signal and its recording into rocks. Nevertheless, the values obtained from paleomagnetic data can be compared to structural estimates. They grossly show that while the expected Greater Asian extent (1,200 km) at 60 Ma can be compared to its shortening (ca. 600–1,000 km), the expected extent great India (ca. 2800 km) greatly exceeds shortening estimates (ca. 1,000 km). This has led to investigate alternative models that are presented in the following.

Reconstructing pre-collisional configuration of Greater India and Greater Asia is therefore a focus of intense research and controversies. For Greater India, many reconstructions have been proposed since Argand's early views (Figure 1.1). As previously discussed, most are based on estimates of the amount of Indian shortening, according to structural data and paleomagnetic reconstructions. Ali and Aitchison (2005) also proposed using the size of Australia to reconstruct the shape of the Indian continent. Indeed, prior to breaking apart in the Early Cretaceous, northern India was facing Western Australia in the eastern Gondwana Continent. Assuming that India and Australia formed a continuous landmass before they broke apart, the shape of Greater India can then be estimated by carefully putting India and Australia back to their Cretaceous position, and measuring the distance separating them. The size and shape of Greater India was also estimated using seismic tomographic imaging of the crust and mantle that enables us to identify – expressed as positive velocity anomalies – the shape and position of dense lithospheric slab remnants that have been subducted into the mantle (e.g. van der Voo et al. 1999; Kosarev et al. 1999). These geophysical methods have enabled us to track precisely the position of the Indian continental lithosphere currently underplated beneath Asia, as well as lithospheric slabs deeper in the mantle that are interpreted to result from their successive breakoff during the collision (Kind et al. 2002; Nábělek et al. 2009; Chen et al. 2015, and Volume 1 – Chapter 4).

### 1.3.2. *Alternative collision configurations*

With new age and convergence constraints on the collision, simple continental collision models would require very large Greater India and Asia and thus involve the massive removal of continental lithosphere by the lateral extrusion of continental blocks, wholesale continental subduction and very high estimates of overriding plate shortening that are difficult to account for with existing geological and geophysical data (e.g. Molnar and Tapponnier 1995; Guillot et al. 2003; Replumaz and Tapponnier 2003; Cogné et al. 2013). Several models have thus more recently been proposed to account for the mismatch between large convergence and smaller shortening in the collision zone since the collision. The first type of models invokes a "Greater India Basin" (Van Hinsbergen et al. 2012; Huang et al. 2015; van Hinsbergen et al. 2019; Yuan et al. 2021). They propose that a terrane drifted off northern India in the Cretaceous, carrying ahead Greater Himalayan sequences, and was separated from mainland India by a wide oceanic basin (Figure 1.4, see Poblete et al. (2021), for a review). This small terrane would have collided with Asia at 60–55 Ma; the Greater India oceanic basin then subducted beneath Asia until the final India–Asia collision in the late Oligocene–early Miocene

(Figure 1.4). Another family of models implies the first collision of India with an intraoceanic arc at 60–55 Ma while subduction is kept along the Asian margin until a much later India–Asia collision. Various versions of this arc collision model have been proposed (Aitchison et al. 2007; Guilmette et al. 2012; Hébert et al. 2012; Gibbons et al. 2015; Jagoutz et al. 2015; Müller et al. 2016; Westerweel et al. 2019). These models invoke collision between India and Asia in the Late Eocene to Miocene, depending on the assumed magnitude of intra-Asian shortening. Finally, it should be noted that a set of "neoclassic" models have been proposed but they imply a continuous subduction of the 2,400 km long Greater Indian continental crust below Asia (Ingalls et al. 2016; Searle 2019).

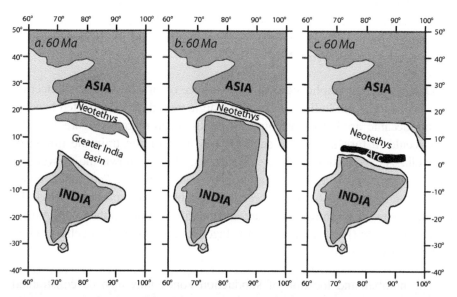

**Figure 1.4.** *Paleogeographic reconstructions according to the three main models proposed for the collision configuration (after Poblete et al. 2021). (a) Greater Indian Basin model, (b) simple continental collision model and (c) arc collision model*

## 1.4. Reconstruction of the Neotethys Ocean closure dynamic

### 1.4.1. *Number of subduction accommodating the Neotethys closure*

Earliest models of Neotethys closure assumed a single subduction zone (Powell and Conaghan 1973; Tapponnier 1981; Honegger et al. 1982; Allègre et al. 1984). The Transhimalaya/Gangdese batholith was recognized as the result of supra-subduction magmatism related to northward Neotethys Ocean beneath the active Asian continental margin starting approximately 250 Ma (Schärer et al. 1984; Ma et al. 2022, for review; Figure 1.5 A). However, detailed studies of the Tethyan ophiolites and associated tectonic mélanges from the Indus-Tsangpo Suture Zone suggested a more complex framework (Aitchison et al. 2000; Mahéo et al. 2000; Hébert et al. 2012, see Volume 2 – Chapter 2 for details). This led to proposing the existence of an intraoceanic subduction, south of the Asian active margin. Such geodynamic setting is also consistent with the tomographic high-velocity anomalies associated with detached oceanic slabs (Spakman 1986; van der Voo et al. 1999; Replumaz et al. 2004, 2013; Hafkenscheid et al. 2006). Based on this observation and interpretation, a range of Neotethys closure models can then be developed assuming a more or less important Greater India and/or the existence of the Greater India Basin (Figure 1.5).

One of the main implications of the existence of an intraoceanic subduction zone is that the north Indian continental margin would have subducted beneath an intraoceanic arc before reaching the south Asian margin (Figure 1.5 B1). This would in turn imply that such proxies used to constrain the timing of the India–Asia collision such as UHP rocks, slowing down the convergent rate, or changing of detrital sediment provenance, may *in fine* rather record an arc–continent collision. Recent reconstructions actually proposed that India collided first with an intraoceanic arc at about 55–60 Ma and then with Asia at about 40–50 Ma (Bouilhol et al. 2013; Jagoutz et al. 2015; Bhattacharya et al. 2020; Martin et al. 2020).

The number and location of subduction zones may also be constrained by seismic tomography. Especially, assuming that slab sink almost vertically Parsons et al. (2020) used tomographic anomalies to reconstruct past subduction locations and their geometries. This study confirmed the presence of two subduction zones and suggests an early collision between intraoceanic arc and the Indian continental margin. At about 60 Ma, the first subduction zone was active along the southern Asia continental margin approximately 20°N, while a second one was active close to the equator. However, these data do not distinguish between the Greater India Basin or wide Greater India models.

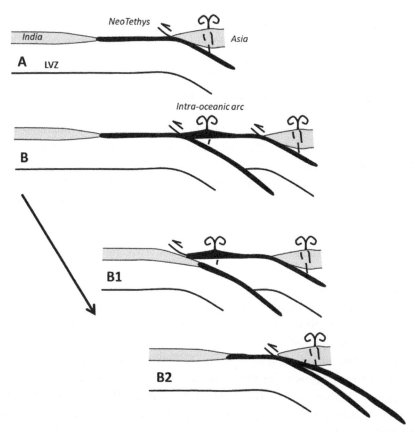

**Figure 1.5.** *Cross-sectional view of the India–Asia convergent zone during the Neotethys closure. A. Single subduction model, B. Double subduction model. B1: Early obduction of the intraoceanic arc on the India margin, B2: Subduction and accretion of the intraoceanic arc in front of the Asian active margin. Horizontal distance not to scale*

## 1.4.2. *Location of the Intraoceanic subduction zone and associated arc*

One of the main questions raised by the two subduction zone models is their relative timing. The intraoceanic arc can either be obducted on top of the Indian continental margin as the subduction below the Asian continent remained active (Figure 1.5B1), or can be accreted to the Asian continental margin before obducting the Indian continent (Figure 1.5B2).

Blueschists within the mélange zone associated with the subduction beneath the Asian margin show a volcanic arc signature (Mahéo et al. 2006). This in turn suggests that an intraoceanic arc was subducted/accreted beneath the Asian margin before collision. Additionally, sedimentary records imply that the mélange zone was active at about 65 Ma.

Such reconstruction seems to contradict with recent paleomagnetic data. (Martin et al. 2020) compare the paleo-latitude evolution of the Greater India with those of the intraoceanic Kohistan–Ladakh arc (see Volume 2 – Chapter 1). This study shows that these areas share a similar latitude about 60 Ma ago close to the equator. At that time, the south Asian margin was located much further north ($\sim$20°N).

An alternative model is then to consider two intraoceanic arcs and one active margin. The Asian active margin would correspond with the Gangdese and Karakorum arcs, as proposed by Jagoutz et al. (2015), the Kohistan–Ladakh arc being part of an intraoceanic arc. Field observations coupled with geochemical analyses in the western Himalaya of India suggest that two sets of ophiolites and the associated mélange zone exist south of Karakorum. One is indeed related to the Kohistan–Ladakh arc (the Dras arc) and one that may account for the presence of another subduction zone south of this latter (Mahéo et al. 2000, 2004; Robertson 2000; Corfield et al. 2001; Catlos et al. 2019). In this configuration, the southernmost arc may be accreted to the Kohistan–Ladakh arc just before the collision with India, accounting for the arc-related late Cretaceous blueschists.

This model seems consistent with the observation and data from western Himalaya but raise the question of its extrapolation eastward and the possibility of non-continuous subduction zone (see Volume 2 – Chapter 2 for details on the ophiolite belts and ophiolite correlation).

### 1.4.3. *Driving forces of the India–Asia convergence during Neotethys closure*

Reconstructions of slab dynamic coupled with kinematic reconstructions allow us to better discuss the forces involved in the convergence between India and Asia. Convergent results from interactions between ridge push, subduction zone forces (slab pull, slab suction and frictions) and mantle underlying plate basal drags (see Faccenna et al. (2021), for details). Recorded changes in India–Asia kinematic was then used to test the relative effects of these forces (e.g. Capitanio et al. 2010; Becker and Faccenna 2011; Cande and Stegman 2011; Copley et al. 2011; van Hinsbergen et al. 2011a; Jagoutz et al. 2015; Jolivet et al. 2018; Pusok and Stegman 2020). Especially, Jagoutz et al. (2015) proposed the modeling of the impact of two subduction zones on convergent rate evolution. This modeling evidences that the first episode of convergent acceleration at about 80 Ma may be related to the decrease of the intraoceanic subduction zone length when this latter is restricted to the India–Asia convergent zone. The following convergent rate slowdown at about 60 Ma is then related to the termination of the intraoceanic subduction zone subsequent to the arc–Indian continent collision and oceanic slab breakoff. Ongoing slowdown is then linked with the final India–Asia collision and continental subduction followed by Neotethys slab breakoff (Chemenda et al. 2000; Kohn and Parkinson 2002; Mahéo et al. 2009) and thickening of the Tibetan plateau (Copley et al. 2011). Continuous India–Asia convergence following the collision is then mostly controlled by mantle drag (Becker and Faccenna 2011). But other processes may also have influenced the India–Asia kinematic and especially the increase of the convergent rate to about 80 Ma. Interaction with mantle plumes such as the Reunion hotspot would also increase the plate velocity (Cande and Stegman 2011; Van Hinsbergen et al. 2011b; Pusok and Stegman 2020). Actually, Pusok and Stegman (2020) proposed that the Reunion hotspot induced both the plate acceleration and the second subduction zone initiation at about 72–66 Ma. Then, the two subduction settings maintain the high convergent rate until the India–arc and India–Asia collisions. In addition, Behr and Becker (2018) suggested that the subduction of equatorial thick sedimentary deposits would significantly decrease the subduction interface viscosity reducing friction and favoring slab pull.

Following collision, slab dynamic significantly changed and several factors such as slab-breakoff, Tibetan plateau buildup or external forcing due to the monsoon may then significantly influence convergence (e.g. Chemenda et al. 2000; Mahéo et al. 2009; Molnar and Stock 2009; Copley et al. 2011; Iaffaldano et al. 2011; Husson et al. 2014; Webb et al. 2017). These aspects will be developed in the next chapters of this book.

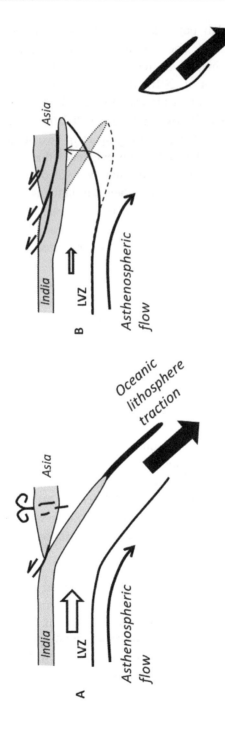

**Figure 1.6.** *Transition from continental subduction (A) to India underplating Asia (B) induced by Neotethys slab detachment*

## 1.5. Conclusion

This review evidences that if the kinematic evolution of the India convergence is relatively well constrained, many points remain discussed such as the size of Greater India or the slab dynamics during Neotethys subduction. Recent coupling between petrologic constrains, plate kinematic and tomographic data appears very promising to better reconstruct the geodynamic evolution of the convergent zone up to the collision.

Such detailed reconstructions are crucial to estimate the initial geometry and organization of the collision zone which may significantly influence both the Himalaya and Tibetan Plateau petrologic evolution, tectonism and surface processes that will be discussed in the next chapters. Notably, the cessation of continental subduction followed by Indian continental lithosphere underplating must have drastically changed the mechanical coupling between Indian and Asian crusts as well as the thermal regime of the convergent zone (Figure 1.6, see Volume 1 – Chapter 2 for details).

Despite a century of continuous research on the India–Asia collision since the work of Emile Argand, the debate on how the collision occurred remains fierce. Following from the initial simple models of a single collision between two continental plates, the acquisition of massive geologic datasets and the evolving geodynamic concepts have led to the development of new models involving several collisions of intervening microplates. Deciphering between these competing models in the years to come will be the focus of exciting research and certainly lead to major breakthroughs in understanding geodynamic processes with major implications for Earth, Climate and Life sciences.

## 1.6. References

Aitchison, J.C., Badengzhu, Davis, A.M., Liu, J., Luo, H., Malpas, J.G., McDermid, I.R.C., Wu, H., Ziabrev, S.V., Zhou, M. (2000). Remnants of a Cretaceous intra-oceanic subduction system within the Yarlung–Zangbo suture (southern Tibet). *Earth and Planetary Science Letters*, 183(1–2), 231–244.

Aitchison, J.C., Ali, J.R., Davis, A.M. (2007). When and where did India and Asia collide? *Journal of Geophysical Research: Solid Earth*, 112, B05423, doi:10.1029/2006JB004706.

Ali, J.R. and Aitchison, J.C. (2005). Greater India. *Earth-Science Reviews*, 72(3–4), 169–188.

Ali, J.R. and Aitchison, J.C. (2008). Gondwana to Asia: Plate tectonics, paleogeography and the biological connectivity of the Indian sub-continent from the Middle Jurassic through latest Eocene (166–35 Ma). *Earth-Science Reviews*, 88(3), 145–166.

Allègre, C.J., Courtillot, V., Tapponnier, P., Hirn, A., Mattauer, M., Coulon, C., Jaeger, J.J., Achache, J., Schärer, U., Marcoux, J. et al. (1984). Structure and evolution of the Himalaya–Tibet orogenic belt. *Nature*, 307(5946), 17–22.

Argand, E. (1924). La tectonique de l'Asie. *Congrés géologique international (XIIIe session)*, Brussels, 1992.

Beck, R.A., Burbank, D.W., Sercombe, W.J., Riley, G.W., Barndt, J.K., Berry, J.R., Afzal, J., Khan, A.M., Jurgen, H., Metje, J. et al. (1995). Stratigraphic evidence for an early collision between northwest India and Asia. *Nature*, 373(6509), 55–58.

Beck, R.A., Burbank, D.W., Sercombe, W.J., Khan, A.M., Lawrence, R.D. (1996). Late Cretaceous ophiolite obduction and Paleocene India-Asia collision in the westernmost Himalaya. *Geodinamica Acta*, 9(2–3), 114–144.

Becker, T.W. and Faccenna, C. (2011). Mantle conveyor beneath the Tethyan collisional belt. *Earth and Planetary Science Letters*, 310(3), 453–461.

Behr, W.M. and Becker, T.W. (2018). Sediment control on subduction plate speeds. *Earth and Planetary Science Letters*, 502, 166–173.

Besse, J. and Courtillot, V. (2002). Apparent and true polar wander and the geometry of the geomagnetic field over the last 200 Myr. *Journal of Geophysical Research: Solid Earth*, 107(B11), EPM 6-1–EPM 6-31.

Bhattacharya, G., Robinson, D.M., Wielicki, M.M. (2020). Detrital zircon provenance of the Indus Group, Ladakh, NW India: Implications for the timing of the India-Asia collision and other syn-orogenic processes. *GSA Bulletin*, 133(5–6), 1007–1020.

Bouilhol, P., Jagoutz, O., Hanchar, J.M., Dudas, F.O. (2013). Dating the India–Eurasia collision through arc magmatic records. *Earth and Planetary Science Letters*, 366, 163–175.

Butler, R.F. (1992). *Paleomagnetism: Magnetic Domains to Geologic Terranes*. Blackwell Scientific Publications, Hoboken.

Cande, S.C. and Stegman, D.R. (2011). Indian and African plate motions driven by the push force of the Réunion plume head. *Nature*, 475(7354), 47–52.

Capitanio, F.A., Morra, G., Goes, S., Weinberg, R.F., Moresi, L. (2010). India–Asia convergence driven by the subduction of the Greater Indian continent. *Nature Geoscience*, 3(2), 136–139.

Capitanio, F.A., Replumaz, A., Riel, N. (2015). Reconciling subduction dynamics during Tethys closure with large-scale Asian tectonics: Insights from numerical modeling. *Geochemistry, Geophysics, Geosystems*, 16(3), 962–982.

Catlos, E.J., Pease, E.C., Dygert, N., Brookfield, M., Schwarz, W.H., Bhutani, R., Pande, K., Schmitt, A.K. (2019). Nature, age and emplacement of the Spongtang ophiolite, Ladakh, NW India. *Journal of the Geological Society*, 176(2), 284–305.

Chatterjee, S. and Scotese, C. (2010). The Wandering Indian Plate and its changing biogeography during the Late Cretaceous-Early Tertiary Period. In *New Aspects of Mesozoic Biodiversity*, Bandyopadhyay, S. (ed.). Springer, Berlin, Heidelberg.

Chemenda, A.I., Burg, J.-P., Mattauer, M. (2000). Evolutionary model of the Himalaya–Tibet system: Geopoem: Based on new modelling, geological and geophysical data. *Earth and Planetary Science Letters*, 174(3), 397–409.

Chen, Y., Courtillot, V., Cogné, J.-P., Besse, J., Yang, Z., Enkin, R. (1993). The configuration of Asia prior to the collision of India: Cretaceous paleomagnetic constraints. *Journal of Geophysical Research: Solid Earth*, 98(B12), 21927–21941.

Chen, Y., Li, W., Yuan, X., Badal, J., Teng, J. (2015). Tearing of the Indian lithospheric slab beneath southern Tibet revealed by SKS-wave splitting measurements. *Earth and Planetary Science Letters*, 413, 13–24.

Chung, S.-L., Chu, M.-F., Zhang, Y., Xie, Y., Lo, C.-H., Lee, T.-Y., Lan, C.-Y., Li, X., Zhang, Q., Wang, Y. (2005). Tibetan tectonic evolution inferred from spatial and temporal variations in post-collisional magmatism. *Earth-Science Reviews*, 68(3), 173–196.

Cogné, J.-P., Besse, J., Chen, Y., Hankard, F. (2013). A new Late Cretaceous to Present APWP for Asia and its implications for paleomagnetic shallow inclinations in Central Asia and Cenozoic Eurasian plate deformation. *Geophysical Journal International*, 192(3), 1000–1024.

Copley, A., Avouac, J.-P., Wernicke, B.P. (2011). Evidence for mechanical coupling and strong Indian lower crust beneath southern Tibet. *Nature*, 472(7341), 79–81.

Corfield, R.I., Searle, M.P., Pedersen, R.B. (2001). Tectonic setting, origin, and obduction history of the Spontang Ophiolite, Ladakh Himalaya, NW India. *The Journal of Geology*, 109(6), 715–736.

Cox, A. and Hart, R.B. (1986). *Plate Tectonics: How it Works*. Blackwell Scientific Publication, Oxford.

Critelli, S. and Garzanti, E. (1994). Provenance of the Lower Tertiary Murree redbeds (Hazara-Kashmir Syntaxis, Pakistan) and initial rising of the Himalayas. *Sedimentary Geology*, 89(3), 265–284.

DeCelles, P.G., Gehrels, G.E., Quade, J., Ojha, T.P., Kapp, P.A., Upreti, B.N. (1998). Neogene foreland basin deposits, erosional unroofing, and the kinematic history of the Himalayan fold-thrust belt, western Nepal. *GSA Bulletin*, 110(1), 2–21.

DeCelles, P.G., Robinson, D.M., Zandt, G. (2002). Implications of shortening in the Himalayan fold-thrust belt for uplift of the Tibetan Plateau. *Tectonics*, 21(6), 12-1–12-25.

DeCelles, P.G., Kapp, P., Gehrels, G.E., Ding, L. (2014). Paleocene–Eocene foreland basin evolution in the Himalaya of southern Tibet and Nepal: Implications for the age of initial India-Asia collision. *Tectonics*, 33(5), 824–849.

Dewey, J.F. and Bird, J.M. (1970). Mountain belts and the new global tectonics. *Journal of Geophysical Research*, 75(14), 2625–2647.

Dewey, J.F., Shackleton, R.M., Chengfa, C., Yiyin, S. (1988). The tectonic evolution of the Tibetan Plateau. *Philosophical Transactions of the Royal Society of London. Series A, Mathematical and Physical Sciences*, 327(1594), 379–413.

Dupont-Nivet, G., Lippert, P.C., Van Hinsbergen, D.J.J., Meijers, M.J.M., Kapp, P. (2010). Palaeolatitude and age of the Indo–Asia collision: Palaeomagnetic constraints. *Geophysical Journal International*, 182(3), 1189–1198.

Faccenna, C., Becker, T.W., Holt, A.F., Brun, J.P. (2021). Mountain building, mantle convection, and supercontinents: Holmes (1931) revisited. *Earth and Planetary Science Letters*, 564, 116905.

Gansser, A. (1964). *Geology of the Himalayas*. Interscience Publishers, Hoboken.

Garzanti, E. (2019). The Himalayan Foreland Basin from collision onset to the present: A sedimentary–petrology perspective. *Geological Society, London, Special Publications*, 483(1), 65–122.

Garzanti, E., Baud, A., Mascle, G. (1987). Sedimentary record of the northward flight of India and its collision with Eurasia (Ladakh Himalaya, India). *Geodinamica Acta*, 1(4–5), 297–312.

Garzanti, E., Limonta, M., Vezzoli, G., An, W., Wang, J., Hu, X. (2018). Petrology and multimineral fingerprinting of modern sand generated from a dissected magmatic arc (Lhasa River, Tibet). In *Tectonics, Sedimentary Basins, and Provenance: A Celebration of William R. Dickinson's Career*, Ingersoll, R.V., Lawton, T.F., Graham, S.A. (eds). Geological Society of America, Boulder.

Gibbons, A.D., Zahirovic, S., Müller, R.D., Whittaker, J.M., Yatheesh, V. (2015). A tectonic model reconciling evidence for the collisions between India, Eurasia and intra-oceanic arcs of the central-eastern Tethys. *Gondwana Research*, 28(2), 451–492.

Guillot, S., Garzanti, E., Baratoux, D., Marquer, D., Mahéo, G., de Sigoyer, J. (2003). Reconstructing the total shortening history of the NW Himalaya. *Geochemistry, Geophysics, Geosystems*, 4(7), 1064, doi:10.1029/2002GC000484, 2003.

Guillot, S., Mahéo, G., de Sigoyer, J., Hattori, K.H., Pêcher, A. (2008). Tethyan and Indian subduction viewed from the Himalayan high- to ultrahigh-pressure metamorphic rocks. *Tectonophysics*, 451(1), 225–241.

Guilmette, C., Hébert, R., Dostal, J., Indares, A., Ullrich, T., Bédard, E., Wang, C. (2012). Discovery of a dismembered metamorphic sole in the Saga ophiolitic mélange, South Tibet: Assessing an Early Cretaceous disruption of the Neo-Tethyan supra-subduction zone and consequences on basin closing. *Gondwana Research*, 22(2), 398–414.

Hafkenscheid, E., Wortel, M.J.R., Spakman, W. (2006). Subduction history of the Tethyan region derived from seismic tomography and tectonic reconstructions. *Journal of Geophysical Research: Solid Earth*, 111, B08401, doi:10.1029/2005JB003791.

Hébert, R., Bezard, R., Guilmette, C., Dostal, J., Wang, C.S., Liu, Z.F. (2012). The Indus–Yarlung Zangbo ophiolites from Nanga Parbat to Namche Barwa syntaxes, southern Tibet: First synthesis of petrology, geochemistry, and geochronology with incidences on geodynamic reconstructions of Neo-Tethys. *Gondwana Research*, 22(2), 377–397.

Hellinger, S.J. (1981). The uncertainties of finite rotations in plate tectonics. *Journal of Geophysical Research: Solid Earth*, 86(B10), 9312–9318.

Hess, H.H. (1962). History of ocean basins. In *Petrologic Studies*, Engel, A.E.J., James, H.L., Leonard, B.F. (eds). Geological Society of America, Boulder.

van Hinsbergen, D.J.J., Steinberger, B., Doubrovine, P.V., Gassmöller, R. (2011a). Acceleration and deceleration of India-Asia convergence since the Cretaceous: Roles of mantle plumes and continental collision. *Journal of Geophysical Research: Solid Earth*, 116, B06101, doi:10.1029/2010JB008051.

van Hinsbergen, D.J.J., Kapp, P., Dupont-Nivet, G., Lippert, P.C., DeCelles, P.G., Torsvik, T.H. (2011b). Restoration of Cenozoic deformation in Asia and the size of Greater India. *Tectonics*, 30(5).

van Hinsbergen, D.J.J., Lippert, P.C., Dupont-Nivet, G., McQuarrie, N., Doubrovine, P.V., Spakman, W., Torsvik, T.H. (2012). Greater India Basin hypothesis and a two-stage Cenozoic collision between India and Asia. *Proceedings of the National Academy of Sciences*, 109(20), 7659–7664.

van Hinsbergen, D.J.J., Lippert, P.C., Li, S., Huang, W., Advokaat, E.L., Spakman, W. (2019). Reconstructing Greater India: Paleogeographic, kinematic, and geodynamic perspectives. *Tectonophysics*, 760, 69–94.

van Hinsbergen, D.J.J., Steinberger, B., Guilmette, C., Maffione, M., Gürer, D., Peters, K., Plunder, A., McPhee, P.J., Gaina, C., Advokaat, E.L. et al. (2021). A record of plume-induced plate rotation triggering subduction initiation. *Nature Geoscience*, 14(8), 626–630.

Holmes, A. (1965). *Principles of Physical Geology*. Nelson, London.

Honegger, K., Dietrich, V., Frank, W., Gansser, A., Thöni, M., Trommsdorff, V. (1982). Magmatism and metamorphism in the Ladakh Himalayas (the Indus-Tsangpo suture zone). *Earth and Planetary Science Letters*, 60(2), 253–292.

Hu, X., Garzanti, E., Wang, J., Huang, W., An, W., Webb, A. (2016). The timing of India-Asia collision onset – Facts, theories, controversies. *Earth-Science Reviews*, 160, 264–299.

Huang, W., Van Hinsbergen, D.J., Lippert, P.C., Guo, Z., Dupont-Nivet, G. (2015). Paleomagnetic tests of tectonic reconstructions of the India-Asia collision zone. *Geophysical Research Letters*, 42(8), 2642–2649.

Husson, L., Bernet, M., Guillot, S., Huyghe, P., Mugnier, J.-L., Replumaz, A., Robert, X., Van der Beek, P. (2014). Dynamic ups and downs of the Himalaya. *Geology*, 42(10), 839–842.

Iaffaldano, G., Husson, L., Bunge, H.-P. (2011). Monsoon speeds up Indian plate motion. *Earth and Planetary Science Letters*, 304(3), 503–510.

Ingalls, M., Rowley, D.B., Currie, B., Colman, A.S. (2016). Large-scale subduction of continental crust implied by India–Asia mass-balance calculation. *Nature Geoscience*, 9(11), 848–853.

Jaeger, J.-J., Courtillot, V., Tapponnier, P. (1989). Paleontological view of the ages of the Deccan Traps, the Cretaceous/Tertiary boundary, and the India-Asia collision. *Geology*, 17(4), 316–319.

Jagoutz, O., Royden, L., Holt, A.F., Becker, T.W. (2015). Anomalously fast convergence of India and Eurasia caused by double subduction. *Nature Geoscience*, 8(6), 475–478.

Jolivet, L., Faccenna, C., Becker, T., Tesauro, M., Sternai, P., Bouilhol, P. (2018). Mantle flow and deforming continents: From India-Asia convergence to Pacific subduction. *Tectonics*, 37(9), 2887–2914.

Kapp, P. and DeCelles, P.G. (2019). Mesozoic–Cenozoic geological evolution of the Himalayan-Tibetan orogen and working tectonic hypotheses. *American Journal of Science*, 319(3), 159–254.

Kind, R., Yuan, X., Saul, J., Nelson, D., Sobolev, S.V., Mechie, J., Zhao, W., Kosarev, G., Ni, J., Achauer, U. et al. (2002). Seismic images of crust and upper mantle beneath Tibet: Evidence for Eurasian Plate subduction. *Science*, 298(5596), 1219–1221.

Klootwijk, C.T., Gee, J.S., Peirce, J.W., Smith, G.M., McFadden, P.L. (1992). An early India-Asia contact: Paleomagnetic constraints from Ninetyeast Ridge, ODP Leg 121. *Geology*, 20(5), 395–398.

Kohn, M.J. and Parkinson, C.D. (2002). Petrologic case for Eocene slab breakoff during the Indo-Asian collision. *Geology*, 30(7), 591–594.

Kosarev, G., Kind, R., Sobolev, S.V., Yuan, X., Hanka, W., Oreshin, S. (1999). Seismic evidence for a detached Indian Lithospheric Mantle beneath Tibet. *Science*, 283(5406), 1306–1309.

Krause, D.W., Sertich, J.J.W., O'Connor, P.M., Curry Rogers, K., Rogers, R.R. (2019). The Mesozoic biogeographic history of Gondwanan terrestrial vertebrates: Insights from Madagascar's Fossil Record. *Annual Review of Earth and Planetary Sciences*, 47(1), 519–553.

Ma, X., Attia, S., Cawood, T., Cao, W., Xu, Z., Li, H. (2022). Arc tempos of the Gangdese batholith, southern Tibet. *Journal of Geodynamics*, 149, 101897.

Mahéo, G., Bertrand, H., Guillot, S., Mascle, G., Pêcher, A., Picard, C., Sigoyer, J.D. (2000). Témoins d'un arc immature téthysien dans les ophiolites du Sud Ladakh (NW Himalaya, Inde). *Comptes Rendus de l'Académie Des Sciences – Series IIA – Earth and Planetary Science*, 330(4), 289–295.

Mahéo, G., Bertrand, H., Guillot, S., Villa, I.M., Keller, F., Capiez, P. (2004). The South Ladakh ophiolites (NW Himalaya, India): An intra-oceanic tholeiitic arc origin with implication for the closure of the Neo-Tethys. *Chemical Geology*, 203(3), 273–303.

Mahéo, G., Fayoux, X., Guillot, S., Garzanti, E., Capiez, P., Mascle, G. (2006). Relicts of an intra-oceanic arc in the Sapi-Shergol mélange zone (Ladakh, NW Himalaya, India): Implications for the closure of the Neo-Tethys Ocean. *Journal of Asian Earth Sciences*, 26(6), 695–707.

Mahéo, G., Blichert-Toft, J., Pin, C., Guillot, S., Pêcher, A. (2009). Partial melting of mantle and crustal sources beneath South Karakorum, Pakistan: Implications for the Miocene geodynamic evolution of the India–Asia Convergence Zone. *Journal of Petrology*, 50(3), 427–449.

Martin, C.R., Jagoutz, O., Upadhyay, R., Royden, L.H., Eddy, M.P., Bailey, E., Nichols, C.I.O., Weiss, B.P. (2020). Paleocene latitude of the Kohistan–Ladakh arc indicates multistage India–Eurasia collision. *Proceedings of the National Academy of Sciences*, 117(47), 29487–29494.

Molnar, P. and Stock, J.M. (2009). Slowing of India's convergence with Eurasia since 20 Ma and its implications for Tibetan mantle dynamics. *Tectonics*, 28(3), doi:10.1029/2008TC002271.

Molnar, P. and Tapponnier, P. (1975). Cenozoic tectonics of Asia: Effects of a continental collision. *Science*, 189(4201), 419–426.

Müller, R.D., Seton, M., Zahirovic, S., Williams, S.E., Matthews, K.J., Wright, N.M., Shephard, G.E., Maloney, K.T., Barnett-Moore, N., Hosseinpour, et al. (2016). Ocean basin evolution and global-scale plate reorganization events since Pangea breakup. *Annual Review of Earth and Planetary Sciences*, 44(1), 107–138.

Müller, R.D., Cannon, J., Qin, X., Watson, R.J., Gurnis, M., Williams, S., Pfaffelmoser, T., Seton, M., Russell, S.H.J., Zahirovic, S. (2018). GPlates: Building a virtual earth through deep time. *Geochemistry, Geophysics, Geosystems*, 19(7), 2243–2261.

Nábĕlek, J., Hetényi, G., Vergne, J., Sapkota, S., Kafle, B., Jiang, M., Su, H., Chen, J., Huang, B.S. (2009). Underplating in the Himalaya-Tibet collision zone revealed by the Hi-CLIMB experiment. *Science*, 325(5946), 1371–1374.

Najman, Y., Jenks, D., Godin, L., Boudagher-Fadel, M., Millar, I., Garzanti, E., Horstwood, M., Bracciali, L. (2017). The Tethyan Himalayan detrital record shows that India–Asia terminal collision occurred by 54 Ma in the Western Himalaya. *Earth and Planetary Science Letters*, 459, 301–310.

Parsons, A.J., Hosseini, K., Palin, R.M., Sigloch, K. (2020). Geological, geophysical and plate kinematic constraints for models of the India-Asia collision and the post-Triassic central Tethys oceans. *Earth-Science Reviews*, 208, 103084.

Patriat, P. and Achache, J. (1984). India–Eurasia collision chronology has implications for crustal shortening and driving mechanism of plates. *Nature*, 311(5987), 615–621.

Patzelt, A., Li, H., Wang, J., Appel, E. (1996). Palaeomagnetism of Cretaceous to Tertiary sediments from southern Tibet: Evidence for the extent of the northern margin of India prior to the collision with Eurasia. *Tectonophysics*, 259(4), 259–284.

Poblete, F., Dupont-Nivet, G., Licht, A., van Hinsbergen, D.J.J., Roperch, P., Mihalynuk, M.G., Johnston, S.T., Guillocheau, F., Baby, G., Fluteau, F. et al. (2021). Towards interactive global paleogeographic maps, new reconstructions at 60, 40 and 20 Ma. *Earth-Science Reviews*, 214, 103508.

Powell, C.McA. and Conaghan, P.J. (1973). Plate tectonics and the Himalayas. *Earth and Planetary Science Letters*, 20(1), 1–12.

Pozzi, J.-P., Westphal, M., Xiu Zhou, Y., Sheng Xing, L., Yao Chen, X. (1982). Position of the Lhasa block, South Tibet, during the late Cretaceous. *Nature*, 297(5864), 319–321.

Pusok, A.E. and Stegman, D.R. (2020). The convergence history of India-Eurasia records multiple subduction dynamics processes. *Science Advances*, 6(19).

Replumaz, A. and Tapponnier, P. (2003). Reconstruction of the deformed collision zone between India and Asia by backward motion of lithospheric blocks. *Journal of Geophysical Research: Solid Earth*, 08(B6), 2285, doi:10.1029/2001JB000661.

Replumaz, A., Kárason, H., van der Hilst, R.D., Besse, J., Tapponnier, P. (2004). 4-D evolution of SE Asia's mantle from geological reconstructions and seismic tomography. *Earth and Planetary Science Letters*, 221(1), 103–115.

Replumaz, A., Negredo, A.M., Guillot, S., Villaseñor, A. (2010). Multiple episodes of continental subduction during India/Asia convergence: Insight from seismic tomography and tectonic reconstruction. *Tectonophysics*, 483(1), 125–134.

Replumaz, A., Guillot, S., Villaseñor, A., Negredo, A.M. (2013). Amount of Asian lithospheric mantle subducted during the India/Asia collision. *Gondwana Research*, 24(3), 936–945.

Robertson, A.H.F. (2000). Formation of mélanges in the Indus Suture Zone, Ladakh Himalaya by successive subduction-related, collisional and post-collisional processes during Late Mesozoic-Late Tertiary time. *Geological Society, London, Special Publications*, 170(1), 333–374.

Rowley, D.B. (1996). Age of initiation of collision between India and Asia: A review of stratigraphic data. *Earth and Planetary Science Letters*, 145(1), 1–13.

Rowley, D.B. (2019). Comparing paleomagnetic study means with apparent wander paths: A case study and paleomagnetic test of the Greater India Versus Greater Indian Basin hypotheses. *Tectonics*, 38(2), 722–740.

Royer, J.-Y., Gordon, R.G., Horner-Johnson, B.C. (2006). Motion of Nubia relative to Antarctica since 11 Ma: Implications for Nubia-Somalia, Pacific–North America, and India-Eurasia motion. *Geology*, 34(6), 501.

Schärer, U., Xu, R.-H., Allègre, C.J. (1984). UPb geochronology of Gangdese (Transhimalaya) plutonism in the Lhasa-Xigaze region, Tibet. *Earth and Planetary Science Letters*, 69(2), 311–320.

Schellart, W.P., Chen, Z., Strak, V., Duarte, J.C., Rosas, F.M. (2019). Pacific subduction control on Asian continental deformation including Tibetan extension and eastward extrusion tectonics. *Nature Communications*, 10(1), 4480.

Searle, M.P. (2019). Timing of subduction initiation, arc formation, ophiolite obduction and India–Asia collision in the Himalaya. *Geological Society, London, Special Publications*, 483(1), 19–37.

Spakman, W. (1986). Subduction beneath Eurasia in connection with the Mesozoic Tethys. *Netherlands Journal of Geosciences / Geologie En Mijnbouw*, 65, 145–153.

Tapponnier, P., Mercier, J.L., Proust, F., Andrieux, J., Armijo, R., Bassoullet, J.P., Brunel, M., Burg, J.P., Colchen, M., Dupré, B. et al. (1981). The Tibetan side of the India–Eurasia collision. *Nature*, 294(5840), 405–410.

Tauxe, L. (2010). *Essentials of Paleomagnetism*. University of California Press.

Torsvik, T.H., van der Voo, R., Preeden, U., MacNiocaill, C., Steinberger, B., Doubrovine, P.V., van Hinsbergen, D.J.J., Domeier, M., Gaina, C., Tohver, E. et al. (2012). Phanerozoic polar wander, palaeogeography and dynamics. *Earth-Science Reviews*, 114(3–4), 325–368.

Vaes, B., Gallo, L.C., van Hinsbergen, D.J.J. (2022). On pole position: Causes of dispersion of the paleomagnetic poles behind apparent polar wander paths. *Journal of Geophysical Research: Solid Earth*, 127, e2022JB023953, https://doi.org/10.1029/2022JB023953.

Vine, F.J. and Matthews, D.H. (1963). Magnetic anomalies over oceanic ridges. *Nature*, 199(4897), 947–949.

van der Voo, R., Spakman, W., Bijwaard, H. (1999). Tethyan subducted slabs under India. *Earth and Planetary Science Letters*, 171(1), 7–20.

Webb, A.A.G., Guo, H., Clift, P.D., Husson, L., Müller, T., Costantino, D., Yin, A., Xu, Z., Cao, H., Wang, Q. (2017). The Himalaya in 3D: Slab dynamics controlled mountain building and monsoon intensification. *Lithosphere*, 9(4), 637–651.

Wegener, A. (1929). *Die Entstehung der Kontinente und Ozeane*. Friedr. Vieweg Sohn Akt. Ges., Braunschweig.

Westerweel, J., Roperch, P., Licht, A., Dupont-Nivet, G., Win, Z., Poblete, F., Ruffet, G., Swe, H.H., Thi, M.K., Aung, D.W. (2019). Burma Terrane part of the Trans-Tethyan arc during collision with India according to palaeomagnetic data. *Nature Geoscience*, 12(10), 863–868.

Yuan, J., Yang, Z., Deng, C., Krijgsman, W., Hu, X., Li, S., Shen, Z., Qin, H., An, W., He, H. et al. (2021). Rapid drift of the Tethyan Himalaya terrane before two-stage India-Asia collision. *National Science Review*, 8(7), nwaa173.

Zhu, D.-C., Wang, Q., Zhao, Z.-D., Chung, S.-L., Cawood, P.A., Niu, Y., Liu, S.-A., Wu, F.-Y., Mo, X.-X. (2015). Magmatic record of India-Asia collision. *Scientific Reports*, 5(1), 14289.

# 2

# Building the Tibetan Plateau During the Collision Between the India and Asia Plates

Anne REPLUMAZ[1], Cécile LASSERRE[2], Stéphane GUILLOT[1],
Marie-Luce CHEVALIER[3], Fabio A. CAPITANIO[4],
Francesca FUNICIELLO[5], Fanny GOUSSIN[1]
and Shiguang WANG[3]

[1]*University of Grenoble Alpes, France*
[2]*Claude Bernard University, Lyon, France*
[3]*Chinese Academy of Geological Sciences, Beijing, China*
[4]*School of Earth, Atmosphere and Environment, Monash University, Clayton, Australia*
[5]*Department of Science, Roma Tre University, Rome, Italy*

## 2.1. Introduction

The Tibetan Plateau is the highest and widest orogenic plateau on Earth, towering over the world at an average altitude of 4,500 m and over 800 km. The Plateau is more than three times wider than the Andes, more than 10 times wider and almost twice as high as the European Alps and more than three times higher than the Zagros (Figure 2.1). The exceptional height and width of the Plateau have profoundly impacted the climate in this part of the world, playing a crucial role in the emergence of the monsoon, intensifying summer rainfall

*Himalaya, Dynamics of a Giant 1*,
coordinated by Rodolphe CATTIN and Jean-Luc EPARD.
© ISTE Ltd 2023.

in the Indian plains, along the sharp southern and eastern edges of the plateau, while forcing aridity within the plateau interiors (see Volume 3 – Chapter 1). Tibet's growth results from the ongoing collision between the Indian and Asian continents since more than 50 Ma, with India acting as an indenter to shorten Asia, driven by northward subduction and mantle convection (e.g. Patriat and Achache (1984), see Volume 1 – Chapter 1). In the last decades, the prolific acquisition of multi-scale data from geology (see Volume 2), active tectonics, seismology (see Volume 1 – Chapter 4) or geodesy approaches, as well as experimental or numerical modeling, has supported solid inferences on the structure of the Plateau helping constraining its growth. While far from being an exhaustive review of the existing data and models, this chapter presents some recent constraints and models that provide an integrated and improved understanding of the enigmatic plateau building.

First, at the surface level, the GPS velocity-derived strain shows that the present-day deformation of the upper crust localizes dominant active shortening in the northern plateau and dominant strike-slip deformation in the central plateau (e.g. Gan et al. 2007; Wang and Shen 2020). These tectonic styles accommodate both thickening and widening of the plateau during the collision (Meyer et al. 1998). Second, at the lithospheric depth, seismic profiles show the southward subduction of the lithospheric Tibetan mantle in the Pamir and in central Tibet (e.g. Zhao et al. 2011; Schneider et al. 2013), opposite to the northward Indian subduction beneath the Himalayan front (see Volume 1 – Chapter 4). A remnant of ancient continental Tibetan slab subduction occurring in the early stage of collision is now imaged in the lower mantle from global tomography (e.g. Replumaz et al. 2014). These subduction episodes are an important process which control the partial accommodation of the northward motions of India during the collision at the lithospheric level, while the remainder acted as an indenter onto the Asian continent. Third, the emplacement of lavas with an arc-like chemical signature shows that the past subduction of the Tibetan lithosphere occurred between $\sim$45 and 30 Ma, along one of the sutures crossing central Tibet, the Jinsha suture (e.g. Spurlin et al. 2005). Petrological analyses of the lavas also show that the long-lasting accretion of the Asian continent before the collision has profoundly modified the rheology of the deep Tibetan lithosphere south of the Jinsha suture, reflected in its modified seismologic velocity signature (Goussin et al. 2020). Such metasomatized mantle is significantly weaker than a normal mantle, favoring strong thickening of the plateau's interior.

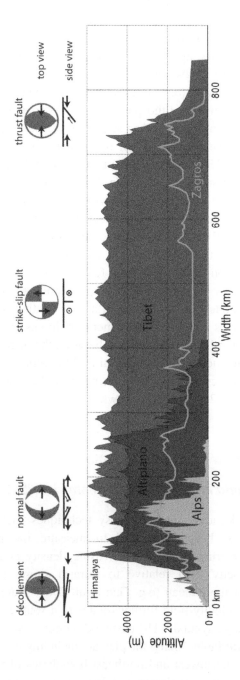

**Figure 2.1.** *Topographic profiles across some mountain ranges of the Earth (modified from Vanderhaeghe 2012). On top, various types of focal mechanisms observed across Tibet. For a color version of this figure, see www.iste.co.uk/cattin/himalaya1.zip*

Despite the growing knowledge on the plateau's deep and shallow structure, the formation of the plateau and its controls remain still unknown. First, while the path and velocity of the Indian plate in the last $\sim$100 Ma are well known since $\sim$40 years ago (e.g. Patriat and Achache 1984), the forces driving India so fast (between 3 and 5 cm/yr) for so long (more than 50 Ma) remain uncertain (e.g. Becker and Faccenna 2011). Then, why the subduction of continental lithosphere or the Asian upper plate did not act to halt the indentation, as in other segments of the Africa–Eurasia collision, is unknown. In the second part of this chapter, we present some analogue and numerical models at the lithospheric/upper mantle scale to better understand the dynamics of subduction, mantle flow and the formation of the plateau as a result. Analogue models have shown that the subduction of the continental lithosphere, less dense than the mantle, is possible in a compressive context (Replumaz et al. 2016). This emphasizes the role of far-field forces acting upon the Indian continent, likely due to mantle drag or the pull from neighboring subduction zones in nature, forcing collision, continental lithosphere underthrusting or subduction and compression in the upper plate. Furthermore, numerical models of subduction of a plate hosting both oceanic and continental lithospheres allow an assessment of forces, revealing the signature of plate-wide lateral stress distribution between the ongoing oceanic and stalled continental subduction, and the ensuing differential trench retreat and advance, respectively, resulting in both extrusion and thickening in the upper plate (Capitanio et al. 2015), similar to what is observed in Tibet in the present day.

## 2.2. Present-day Tibetan crustal deformation

### 2.2.1. *GPS velocity field and focal mechanisms in Tibet*

Over the past three decades, space geodesy techniques such as global positioning system (GPS) have been used to measure the present-day deformation of the upper crust, with ever-increasing density and precision. In Tibet, the GPS velocity field relative to Eurasia shows a dominant north-eastward motion of the plateau (e.g. Gan et al. 2007; Wang and Shen 2020) (Figure 2.2). The velocity field further rotates to a southeast direction around the eastern Himalaya syntaxis. The eastward motion is accommodated by several major left-lateral faults slicing the plateau, including the Altyn Tagh and Haiyuan faults at the northwest and northeast boundaries of the plateau, respectively, and the Kunlun, Xianshuihe and Jiali faults in the central and eastern parts of the plateau (Figure 2.2). Extension is also observed mostly across grabens in southern-central Tibet. To the north, the velocity amplitude drops abruptly on the edge of the plateau across the thrust faults of the Qilian Shan thrust belt between the Kunlun fault and the Gobi Ala Shan platform considered as stable, showing that active shortening occurs there. Another major drop in the amplitude of the GPS velocity occurs between the Indian continent and the southern plateau, highlighting the ongoing shortening across the Himalayan range, which results in building the highest mountain belt of the world (Figure 2.2).

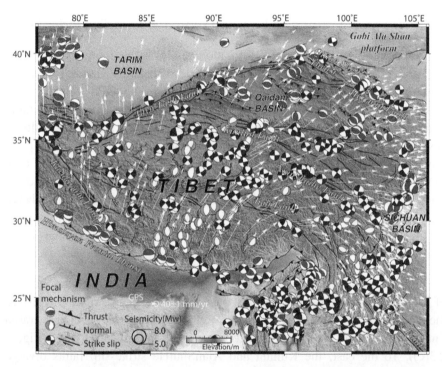

**Figure 2.2.** *GPS velocity field over the Tibetan Plateau considering Eurasia as fixed (Wang and Shen 2020), with earthquake focal mechanisms (Mw>5.0, 1976–2021, https://www.globalcmt.org). For a color version of this figure, see www.iste.co.uk/ cattin/himalaya1.zip*

The distribution of focal mechanisms clearly illustrates this peculiarity between the central part of the plateau where strike slip and extension occur, and the northern and southern parts, where shortening occurs (Figure 2.2). North of the Kunlun fault, parallel thrusts thickening the upper crust are observed at the surface, gradually migrating to the north along the Altyn Tagh fault, and branching at depth on an intra-crustal décollement (Meyer et al. (1998), see figure in the next section). Such a coupling between the horizontal and vertical motions yields the thickening of the Tibetan crust and the widening of the plateau during the collision. It also implies that the northward movement of India is transferred to the north of the plateau resulting in an active shortening, while the central part of the plateau has reached a threshold of elevation, for which the widening of the plateau minimizes energy with respect to its thickening.

### 2.2.2. *Surface motions and deformation due to Indian indentation*

A peculiarity of the plateau deformation pattern is the dominance of strike slip and normal faults (Figure 2.2) where the mean elevation is reached (Figure 2.1). Considering central Tibet as fixed, the GPS velocity field (Gan et al. 2007) shows a strong gradient across the Xianshuihe fault, highlighting a strong localization of deformation on the strike-slip faults of eastern Tibet (Figure 2.3(a)). There, the eastern Qiangtang terrane is rapidly moving eastward, between the Xianshuihe fault to the north and the Jiali fault to the southeast, following at the first order a rigid block rotation around the Eastern Himalaya Syntaxis (e.g. Bai et al. 2018). At the scale of the Tibetan Plateau, the strike-slip faulting is compatible with the slip-lines resulting from an indentation in a homogeneous material (Tapponnier and Molnar 1976) (Figure 2.3(b)).

This simple pattern has to be adapted to take into account the heterogeneity of the Tibetan lithosphere, especially due to the presence of rigid cratons. To the west of the collision zone, the Altyn Tagh fault bordering the Tarim craton limits the plateau to the northwest, accommodating its eastward motion (Figure 2.3(c)). To the east, the Kunlun and XianShuiHe strike-slip faults diverge around the Sichuan craton. Furthermore, the southernmost Karakorum–Jiali fault zone (KJFZ) forms a discontinuous en-echelon fault zone, decoupling between the crusts of northern and southern Tibet and allowing central Tibet's Qiangtang terrane, considered as mostly rigid, to move eastward bypassing the Eastern Himalaya Syntaxis (EHS) (Figure 2.3(c)). By contrast, southern Tibet deforms in a divergent manner due to the curved shape of the Himalayan arc, generating extension through rifts confined within the arc curvature (Armijo et al. 1986). However, considering central Tibet as a rigid block does not explain the presence of rifts in the Qiangtang terrane north of the Karakorum–Jiali fault zone (Figure 2.3). The strain rates in the north are only ~10–20% of those located south of the Karakorum–Jiali fault zone (e.g. Chevalier et al. 2020), suggesting that the eastward extrusion of the western and central parts of the Qiangtang terrane is absorbed by significant internal extension distributed on numerous scattered normal faults, so that it can no longer be treated as an integral block (Han et al 2019). This internal deformation may be due to the absence of major strike-slip faults north of western Qiangtang, hence limiting an efficient eastward extrusion (Wang et al. 2021). More in general, extension has been argued to be due to the gravitational collapse of the high plateau, driven by buoyancy forces, generating an outward lower crustal viscous flow from the plateau center (e.g. Clark and Royden 2000). In eastern Tibet, the flow around the Sichuan Basin should generate normal faulting parallel to the edge of the plateau where the topographic gradient is the largest (Copley 2008). Nevertheless, extension is mostly localized in the center of the plateau, far from the edge, and to the east mostly observed at the junction of strike-slip faults, dominating the deformation (Figure 2.2).

## 2.3. Tibetan lithospheric mantle subduction during collision

### 2.3.1. *Imaging ongoing subduction beneath Tibet*

At depth, seismic profiles are used to constrain the deformation of the lower crust and of the lithospheric mantle beneath Tibet (Figure 2.4, see Volume 1 – Chapter 4). Beneath the Pamir, west of the collision zone, the TIPAGE seismic profile shows a clear southward subduction of Pamir lithosphere plunging down to at least 200 km depth, underlined by deep crustal seismicity, revealing the deepest earthquakes observed on Earth beneath a continental crust (Schneider et al. 2013). The receiver function method, enhancing converted S waves from P waves of distant earthquakes impinging on interfaces beneath the recording stations, gives very accurate depths of density contrast interfaces, clearly confirming that the slab is continental, with the lower crust still attached to the lithospheric mantle (Figure 2.4(a)). Therefore, in the Pamir region, the coupling between subduction of the Pamir lithospheric mantle and lower crust with the shortening of the decoupled upper crust, too buoyant to subduct, accommodates the convergence of India (Sobel et al. 2013).

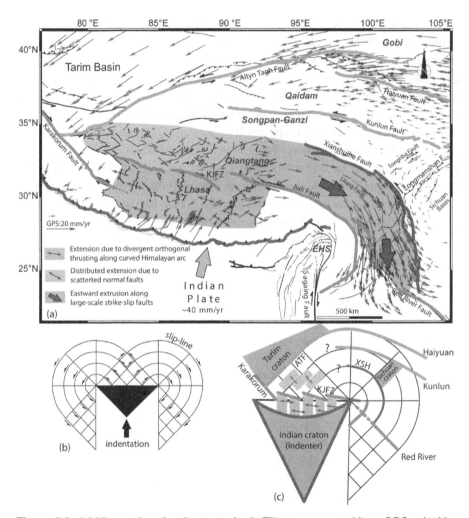

**Figure 2.3.** *(a) Kinematics of active tectonics in Tibet represented from GPS velocities relative to Tibetan Plateau (Gan et al. 2007). (b) Tapponnier and Molnar (1976) slip-lines model predicting the geometry of strike-slip faults resulting from in-plane forces due to collision/indenter. (c) Interpretation of (b) for Tibet (Wang et al. 2021). For a color version of this figure, see www.iste.co.uk/cattin/himalaya1.zip*

Along the INDEPTH seismic profile (e.g. Zhao et al. 2011), crossing southern and central Tibet, a south dipping interface observed with the PS receiver function method has been interpreted as a steep southward subduction of the Asian lithospheric mantle (ALM) beneath the Kunlun fault, extending to at least ∼300 km and flattening below 300 km beneath central Tibet (Figure 2.4(b)). This interpretation is consistent with the thickening of the upper crust north of the Kunlun fault along parallel thrusts branching on the Altyn Tagh strike-slip fault propagating the deformation to the north, while the lower crust attached to the lithospheric mantle subducts southward (Meyer et al. 1998). The coupling of both processes illustrates the complex accommodation of the Asia–India convergence (Figure 2.4(d)). In this case, the lithospheric mantle is considered rigid enough to bend and plunge in the mantle, as observed in Pamir. However, the south dipping interface south of Kunlun is less clearly imaged than that in the Pamir, with no deep crustal seismicity related to such slab (Figure 2.4(c)), nor it is imaged by any method used, such as with the S receiver function method (e.g. Zhao et al. 2011). Furthermore, no other slab is observed between the ALM and the Indian plate (Figure 2.4(b)), so that other interpretations of the structure and evolution of the plateau have been proposed. These range from the homogeneous thickening and convective removal of the thick, unstable lithospheric mantle, to the gravity-driven lower crustal lateral flow to form an orogenic plateau, later dismantled by gravitational collapse (e.g. Vanderhaeghe 2012; Kapp and DeCelles 2019).

## 2.3.2. *Imaging subduction of lithospheric Tibetan mantle during the collision*

The global P-wave tomography images of the mantle below the collision zone have been intensely used during the last 20 years to constrain the deep structure of the Tibetan Plateau at the scale of the entire collision zone (e.g. van der Voo et al. 1999). P-wave tomography shows velocity anomalies in the mantle, where positive anomalies are interpreted as cold lithosphere or slab below the collision zone (e.g. van der Hilst et al. 1997).

Under the assumptions of negligible lateral advection, deeper anomalies are interpreted as older sinking slabs, constraining the subduction history of the Tethys ocean, before the collision, and that of the Indian and Asian lithospheres, during the collision (e.g. Replumaz et al. 2014). In the upper mantle, the slab plunging beneath the Kunlun fault corresponds to a low-amplitude positive anomaly beneath central Tibet (labeled ALM in Figure 2.4).

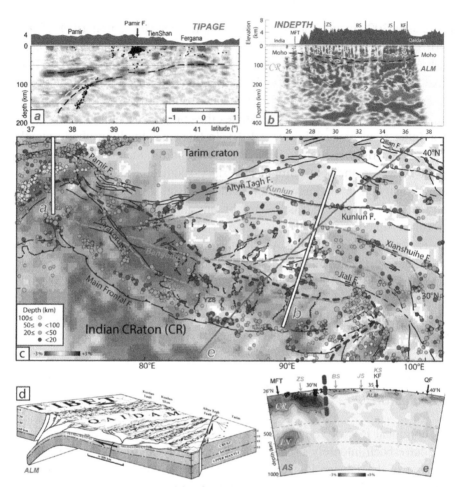

**Figure 2.4.** *Crustal and lithospheric structure beneath Tibet. (a) TIPAGE seismic profile across Pamir (Schneider et al. 2013). (b) INDEPTH seismic profile across Tibet (Zhao et al. 2011). (c) Tomographic section at a 200 km depth (Replumaz et al. 2014), with instrumental seismicity, faults (in black), sutures (in green) and location of the profiles. (d) Schematic 3D block of northern Tibet (Meyer et al. 1998). (e) Tomographic cross-section. Positive anomalies: CR: Indian CRaton, IN: INdia, AS: ASia, ALM: Asian lithospheric mantle. For a color version of this figure, see www.iste.co.uk/cattin/himalaya1.zip*

In the lower mantle, a low-amplitude high wave-speed anomaly (labeled AS) with an elongated geometry of ~3,000 km, running rather parallel to the plate boundary at the onset of the collision, is observed between 1,100 and 900 km-depth in the tomography images (Replumaz et al. 2014). It has been interpreted as a remnant of an episode of Tibetan lithospheric mantle subduction, equivalent to the slab currently observed beneath the Kunlun fault, detached from the continent and sinking into the mantle since then (Figure 2.4). Beneath India, the P-wave tomography cross-sections show a strong positive anomaly (+2-3%), with thick and cold lithospheric mantle of the Indian craton (labeled CR). Such an anomaly extends north of the Tsangpo suture and follows the geometry of the Indian lithosphere, bending south of the suture and underthrusting beneath southern Tibet, as imaged by the Hi-CLIMB profile (Nábělek et al. (2009), see details in Volume 1 – Chapter 4). At 200 km-depth, the amount of underthrusting appears larger to the west. At depths between 900 and 500 km, a prominent anomaly (IN) is observed beneath India, interpreted as a slab of a large portion of the northwestern margin of India (Replumaz et al. 2014).

On the contrary, a low P-wave velocity anomaly observed beneath western central Tibet (Figure 2.4(c) and (e)) has been interpreted as a hotter thermal anomaly, likely due to mantle upwelling replacing lithospheric mantle. This anomaly has been associated with lithosphere delamination or slab retreat under its own weight, leading to lithospheric mantle thinning below Tibet (e.g. Vanderhaeghe 2012; Kapp and DeCelles 2019).

### 2.3.3. *Volcanism in Tibet showing the subduction of lithospheric Asian mantle during the early collision*

Knowledge of the melt properties, melting modalities, and geochemical characteristics of the deepest parts of the Tibetan lithosphere is crucial to constrain its rheological properties and to better understand the deformation mechanisms leading to the plateau building. This has been undertaken by focusing on post-collisional mantle-derived magmatism. The earliest stages of magmatism are found in Eocene volcano-sedimentary basins extending for more than 1,500 km along the Jinsha suture between the Qiangtang and Songpan–Ganze terranes of central and eastern Tibet (Figure 2.5(a)). In the Nangqian basin in eastern Qiangtang, potassic to ultra-potassic calk-alkaline rocks were emplaced at 35–38 Ma during the Cenozoic shortening, attributed to the continental subduction of the Songpan–Ganze lithosphere beneath the Qiangtang terrane along the Jinsha suture (Spurlin et al. 2005).

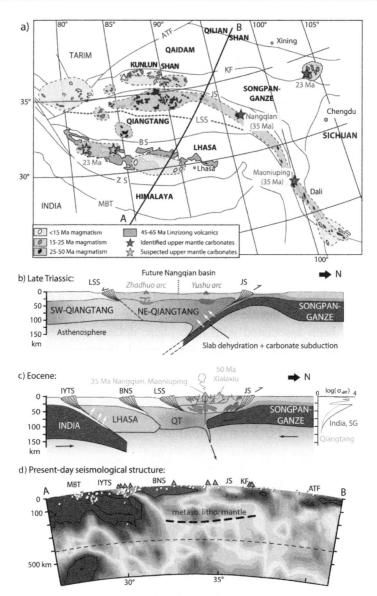

**Figure 2.5.** *Identification of metasomatized lithospheric mantle beneath Central Tibet (modified from Goussin et al. (2020). (a) Map of Cenozoic volcanism. (b) Late Triassic slab dehydration. (c) Eocene compression during the collision. (d) Re-interpretation of the tomographic cross-sectional AB (see location on (a)). For a color version of this figure, see www.iste.co.uk/cattin/himalaya1.zip*

Magmatism occurs mainly as small dykes and sills intruding the mudstone-dominated uppermost basin fill. Folded dykes and growth strata indicate that deformation, magmatism and part of the sedimentation were coeval. A petrological approach has been used to probe the source of Eocene mafic rocks by studying mantle phlogopite xenocrysts and carbonate-bearing ultramafic cumulates preserved in the lavas, identified as a metasomatized lithospheric mantle strongly enriched in volatiles, mainly $H_2O$ and $CO_2$ (Goussin et al. 2020). Such an enrichment occurred prior to the India–Asia collision, during the accretion of the terranes forming the Asian continent, following successive episodes of convergence and subduction that built up the fabric of the Asian lithosphere (Figure 2.5(b)).

Phase equilibrium calculation, rheological mixing model and mineral flow laws allow us to reassess realistic density, seismic signature and strength of a $H_2O$- and $CO_2$-rich metasomatized peridotite mineral assemblage in Tibet. Adding such volatiles to the average mantle composition reduces its density and the seismic wave velocities decrease (Goussin et al. 2020). When applied to the Tibet interiors, the seismic signature of the volatile-rich lithosphere supports the reinterpretation of the tomography cross-sections beneath Tibet (Figure 2.5(d)). Indeed, at 200 km-depth, the negative P-wave velocity anomaly (slow velocity) beneath the eastern Qiangtang and Songpan–Ganze terranes has been first interpreted in terms of elevated temperatures, supporting scenarios of lithospheric mantle removal in this area (e.g. Kapp and DeCelles 2019), but analyses of the lavas reveal that it records a compositional, rather than a thermal, anomaly (Goussin et al. 2020). Such a metasomatized mantle is significantly weaker than a normal mantle to facilitate the thickening of the central Tibet lithosphere in the Eocene, but buoyant enough to prevent its sinking into the deep mantle (Figure 2.5(c)).

## 2.4. Modeling the Tibetan plateau formation during the indentation of the Indian continent into Asia

Continental lithosphere has long been considered too buoyant to subduct on its own, and when following the attached denser oceanic slab into the trench during oceanic closure, break-off should occur to prevent continental subduction (e.g. Wong et al. 1997). However, in the last decade, evidences of such counter-intuitive process have been obtained for the subduction of the Indian and Asian continental lithospheres during the Cenozoic collision, present-day beneath the Pamir and central Tibet (e.g. Zhao et al. 2011; Schneider et al. 2013), and in the early collision time beneath central Tibet (e.g. Replumaz et al. 2014; Goussin et al. 2020). Since then, the processes allowing the subduction of a continental lithosphere have been widely explored, and several models successfully reproduced the observations. Among them, we present here analogue and numerical models focusing on the interactions between the Tibetan and Indian continental lithosphere subduction and the Tibetan upper plate deformation during the collision.

## 2.4.1. *Analogue modeling of the Tibetan lithosphere subduction during the indentation of India*

Analogue modeling employs materials with properties analogous to natural materials to reproduce processes at the spatial and temporal scales accessible in a laboratory. To explore the subduction of the tectonic plates, modelers use thin sheets of silicone putty as analogue for lithospheres, lying on top of the low-viscosity glucose syrup simulating the creeping asthenospheric mantle. The use of silicone, which thickens by pure shear, does not adequately reproduce the complex thickening process of the continental upper crust (Figure 2.4(d)), yet adequately simulates the subduction of the lithosphere, the tectonic regime, whether tensional or compressive, and the overall lithospheric–mantle dynamics. To explore the subduction of the Tibetan lithosphere due to the indentation of the Indian continent, a model has been conducted with two plates, each one composed of two layers (Replumaz et al. 2016). The subducting plate is composed of a silicone layer denser than the mantle representing the oceanic lithosphere (analogue for the dense Tethys ocean subducting before the collision) attached to a silicone layer lighter than the mantle representing the continental indenter (analogue for the buoyant Indian continent). The trailing edge of the indenter is attached to a piston advancing at a constant velocity, representing far-field constraints (Figure 2.6(a)). In the upper plate, the layer closer to the indenter (upper plate), representing the Tibetan lithosphere south of the Jinsha suture, is made of lower-viscosity silicone than the layer farther from it (backwall plate), representing the Tibetan lithosphere north of the suture, made of the same silicone as the indenter, the trailing edge of which is fixed to the wall (Figure 2.6(a)).

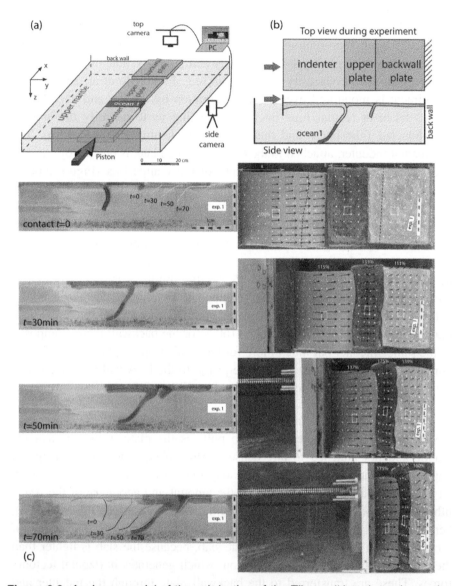

**Figure 2.6.** *Analogue model of the subduction of the Tibetan lithosphere due to the indentation of the Indian continent (modified from Replumaz et al. (2016). (a) 3D model setup and corresponding top and side views of the model. (b) Side and top views of the experiment between contact of the indenter with the upper plate (t=0) and with the backwall plate (t=70 minutes). For a color version of this figure, see www.iste.co.uk/cattin/himalaya1.zip*

The plates are placed in a Plexiglas box (75 x 75 x 25 cm³) on top of 11 cm of glucose syrup, representing the upper mantle (660 km). The experiments are properly scaled for gravity, length, density and viscosity, adapting the velocity of the piston to the viscosity of the syrup (e.g. Funiciello et al. 2003). Each experiment is monitored over its entire duration by the top and lateral view photos taken at regular time intervals, enabling us to quantify the amount of subducted lithosphere and that of plate thickening (Figure 2.6(b)). The continental indenter, attached to the dense oceanic plate, is pushed by a piston and subducts to two-thirds of the depth of the mantle box (Figure 2.6(c)). The convergence is absorbed first by the subduction of the indenting plate, then by both the subduction of the indenter and backwall plate, as well as by the deformation of the three plates. At the beginning of the experiment, the oceanic slab attached to the indenter generates a positive slab pull where the slab dips vertically until it reaches the bottom of the box. Then, it pulls the light indenter down into the mantle overturning the slab and deforming it greatly (Figure 2.6(c)). The vertical component of velocity decreases rapidly before stopping. The thickness increases continuously for the three plates, at maximum ~175% for the indenter, ~160% for the backwall plate, composed of the same silicone, but much more for the weaker upper plate, ~238%. The horizontal velocity decreases from the piston to the backwall as convergence is partitioned into subduction and thickening (Figure 2.6(c)).

Surprisingly, the backwall plate also subducts continuously but slower than the indenter. Indeed, there is no slab pull, as the plate is lighter than the mantle and not attached to any dense oceanic plate. The slab is steep but not vertical, and it reaches approximately one-third of the box depth. Such an experiment adequately reproduces the amount and geometry of the Asian lithosphere subduction episodes inferred during the collision (Figure 2.4(b)), hence correctly reproducing the physics behind such a process. The engine of this subduction is not the weight of the slab, because the slab is lighter than the mantle, but the motion of the piston, which generates horizontal tectonic forces. These are transferred to the backwall plate through the indenter and the upper plate, even if highly deformable, and by the advancing indenter slab through the mantle at shallow depths.

The authors define this process of subduction not driven by the slab pull as "collisional subduction" occurring in a compressional context (Replumaz et al. 2016). The collisional subduction absorbs between 14 and 20% of the convergence, and thus represents an important component of collisional mass balance. The transfer of tectonic forces far from the collision front allows the strong thickening of the upper plate between the Indian and Tibetan continental lithospheres' subductions (Figure 2.6(c)). The weaker metasomatized mantle beneath central Tibet would facilitate such a strong thickening (Goussin et al. 2020).

## 2.4.2. *Numerical modeling of Asian thickening and extrusion during the subduction of a continental–oceanic plate*

Three-dimensional numerical models have been developed to investigate the relationship between subduction dynamics and the large-scale tectonics of continent interiors (Capitanio et al. 2015). These models focus on the impact of the subduction of the Tethys plate hosting continental (India) and the oceanic (Indian Ocean) lithospheres on the tectonics of the Asian continent interior. The models adopted a composite visco-plastic rheology, embedding two deformation modes, Newtonian temperature-independent viscous flow at low stresses and plasticity at larger stresses capture the process of faulting and shearing relevant to the Asian tectonics (Figure 2.3). The upper plate has no density contrast with the mantle and has a free-slip boundary condition on the model's top, that is, no free surface. This choice allows the break-down of topography formation into the isostatic component, due to lithospheric thickness variations, and a dynamic support from the mantle flow. Far-field forces are imposed to the tail of the downgoing plate, ensuing stresses at the convergent margins scaled with ridge push values.

During the initial stage, the oceanic lithosphere sinks into the mantle under the pull of its own negative buoyancy and develops a subduction zone along the margin, driving plate convergence. In the first stage, the margin is stable and deformation is mostly accommodated at the trench, that is, no deformation propagates into the upper plate.

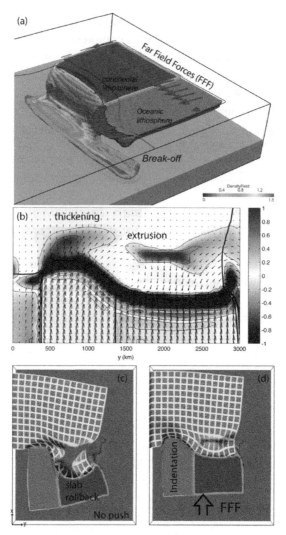

**Figure 2.7.** *Subduction dynamics of an heterogeneous plate (modified from Capitanio et al. (2015). (a) 3D view of the model. (b) Lithospheric thickening (in blue) and thinning (in red) with surface velocity for the model with large far-field forces (FFF). (c) Top view of the model with no push. (d) Top view model with FFF. For a color version of this figure, see www.iste.co.uk/cattin/himalaya1.zip*

In the second stage, the continental lithosphere is entrained in the trench; the density change along the subducted margin forces large stresses at the ocean–continent boundary (OCB), yielding to the detachment of the deep portion of the oceanic slab from the continental lithosphere (Figure 2.7(a)).

The subduction of the continental lithosphere locally reduces the subducting forces, which eventually vanishes upon breakoff, pinning the collision zone. This forces a rearrangement of the convergence accommodation and the differential retreat of the oceanic trench. When far-field forces are added to this system, the incoming plate rate increases, balancing the oceanic trench retreat while the continental trench advances. By adding far-field forces twice the ridge push, the upper plate deformation reproduces tectonic patterns of indentation, with the lateral escape of continental blocks towards the oceanic trench and compression above the continental subduction (Figure 2.7(b)). Such patterns are remarkably similar to the present-day deformation, such as that shown by GPS (Figure 2.2), with the coupling between India's indentation, thickening in Tibet, and eastward extrusion of Southeast Asia. By adding far-field forces only equal to ridge push, the oceanic margin retreats by the same amount of the model with no ridge push while the continental margin's advance, indenting the upper plate but with negligible thickening in the area in front of the indenter (Capitanio et al. 2015).

Therefore, strong far-field forces are necessary to reproduce inferences on the balance between the driving mechanisms of large-scale indentation, lithospheric thickening and faulting in Tibet (Figure 2.7(c)). Far-field forces refer to forces that contribute to plate tectonics but that originate in the deep mantle, remotely from plate boundaries. This reveals that slab pull is not the only driver of plate tectonics and that mantle drag from the underlying mantle convection is also essential to drive plates against each other (e.g. Becker and Faccenna 2011), as well as the large-scale effect of neighboring subduction zones, along wide plate margins (Capitanio et al. 2015). Indeed, the Indian collision is embedded in a much wider convergent system, the Tethyan subduction zone, constraining the force balance at the boundary between India and Eurasia, and driving the northward motion of India at a rate in excess of 50 mm/yr since the beginning of the collision (e.g. Patriat and Achache 1984).

## 2.5. Conclusion

The Tibetan Plateau results from the collision between the Indian and Asian continents. India acts as an indenter shortening and thickening the Asian lithosphere, which ultimately uplifted to form the elevated surface of the Plateau (Figure 2.1). Buoyant continent subduction and slab breakoff since inception, 50 Ma, imply that slab pull forces along the margin are not sufficient to explain the dimension of this vast deformation zone, propagating 100s km away from the plate margin, into the Asian continent interior (Figure 2.2). The forces necessary to drive India so far northward are best understood in a much wider context system, including the deep mantle and the neighboring Indian ocean subduction. At present, the GPS velocity field shows a similar pattern with shortening and lateral extrusion over a vast region (e.g. Gan et al. 2007; Wang and Shen 2020), driven by large strike-slip faults slicing the Tibetan upper crust (Figure 2.3).

Crustal seismic profiles in the Pamir (Schneider et al. 2013) and in central Tibet (Zhao et al. 2011) additionally show complexities in the subduction that do not reconcile with our understanding of subduction-driven collision. Here, the southward subduction of the lithospheric Tibetan mantle, partly absorbing the northward indentation of India, shows evidence of such counter-intuitive process (Figure 2.4). Global tomography suggests anomaly interpreted as the continental Asian slab, now sinking in the lower mantle (Replumaz et al. 2014), with a previous Asian subduction event during early collision (Figure 2.4). Petrological analyses of lavas from central Tibet support the idea that a major subduction event of the Asian lithosphere occurred between $\sim$45 and 30 Ma, and point to a hydrated lithospheric mantle beneath central Tibet (Goussin et al. 2020). Indeed, several continental lithosphere subduction events occurred during the long-lasting accretion of the Asian continent before the collision, hydrating the mantle beneath central and southern Tibet (Figure 2.5). Such hydration is related to a negative seismic velocity anomaly, while ruling out the thermal anomaly previously associated with the delamination of the Tibetan lithospheric mantle may have favored thickening of the Tibetan lithosphere south of the Jinsha during the collision, without requiring excess lithospheric removal.

## 2.6. References

Armijo, R., Tapponnier, P., Mercier, J.L., Han, T.L. (1986). Quaternary extension in southern Tibet: Field observations and tectonic implications. *Journal of Geophysical Research*, 91(B14), 13,803–13,872. doi: 10.1029/JB091iB14p13803.

Bai, M., Chevalier, M.L., Pan, J., Replumaz, A., Leloup, P.H., Métois, M., Li, H. (2018). Southeastward increase of the Late Quaternary slip-rate of the Xianshuihe fault, eastern Tibet. Geodynamic and seismic hazard implications. *Earth and Planetary Science Letters*, 485, 19–31. doi: 10.1016/j.epsl.2017.12.045.

Becker, T.W. and Faccenna, C. (2011). Mantle conveyor beneath the Tethyan collisional belt. *Earth and Planetary Science Letters*, 310, 453–461. doi:10.1016/j.epsl.2011.08.021.

Capitanio, F.A., Replumaz, A., Riel, N. (2015). Reconciling subduction dynamics during Tethys closure with large-scale Asian tectonics: Insights from numerical modeling. *Geochemistry, Geophysics, Geosystems*, 16, 962–982. doi:10.1002/2014GC005660.

Chevalier, M.L., Tapponnier, P., van der Woerd, J., Leloup, P.H., Wang, S., Pan, J., Bai, M., Kali, E., Liu, X., Li., H. (2020). Late Quaternary extension rates across the Northern Half of the Yadong-Gulu rift–implication for East-West extension in Southern Tibet. *Journal of Geophysical Research*, 125. doi:10.1029/2019JB019106.

Clark, M.K. and Royden, L.H. (2000). Topographic ooze: Building the eastern margin of Tibet by lower crustal flow. *Geology*, 28(8), 703–706.

Copley, A. (2008). Kinematics and dynamics of the southeastern margin of the Tibetan Plateau. *Geophysical Journal International*, 174(3), 1081–1100.

Funiciello, F., Faccenna, C., Giardini, D., Regenauer-Lieb, K. (2003). Dynamics of retreating slabs 2: Insights from three-dimensional laboratory experiments. *Journal of Geophysical Research*, 108(B4), 2207. doi:10.1029/2001jb000896.

Gan, W., Zhang, P., Shen, Z., Niu, Z., Wang, M., Wan, Y., Cheng, J. (2007). Present-day crustal motion within the Tibetan Plateau inferred from GPS measurements. *Journal of Geophysical Research*, 112. doi: 10.1029/2005JB004120.

Goussin, F., Riel, N., Cordier, C., Guillot, S., Boulvais, P., Roperch, P., Replumaz, A., Schulmann, K., Dupont-Nivet, G., Rosas, F. et al. (2020). Carbonated inheritance in the Eastern Tibetan lithospheric mantle: Petrological evidences and geodynamic implications. *Geochemistry, Geophysics, Geosystems*, 21(2). doi:10.1029/2019GC008495.

Han, S., Li, H., Pan, J., Lu, H., Zheng, Y., Liu, D., Ge, C. (2019). Co-seismic surface ruptures in Qiangtang Terrane: Insight into Late Cenozoic deformation of central Tibet. *Tectonophysics*, 750, 359–378. doi :10.11016/j.tecto.2018.11.001.

van der Hilst, R.D., Widiyantoro, S., Engdahl, E.R. (1997). Evidence for deep mantle circulation. *Nature*, 386, 578–584.

Kapp, P. and DeCelles, P.G. (2019). Mesozoic–Cenozoic geological evolution of the Himalayan–Tibetan orogen and working tectonic hypotheses. *American Journal of Science*, 319(3), 159–254. doi:10.2475/03.2019.01.

Meyer, B., Tapponnier, P., Bourjot, L., Métivier, F., Gaudemer, Y., Peltzer, G., Shunmin, G., Zhitai, C. (1998). Mechanisms of active crustal thickening in Gansu-Qinghai, and oblique, strike-slip controlled, northeastward growth of the Tibet Plateau. *Geophysical Journal International*, 133, 1–47.

Nábělek, J., Hetényi, G., Vergne, J., Sapkota, S., Kafle, B., Jiang, M., Su, H., Chen, J., Huang, B.S. (2009). Underplating in the Himalaya-Tibet collision zone revealed by the Hi-CLIMB experiment. *Science*, 325(5946), 1371–1374.

Patriat, P. and Achache, J. (1984). India–Eurasia collision chronology has implications for crustal shortening and driving mechanisms of plates. *Nature*, 311, 615–621.

Replumaz, A., Capitanio, F.A., Guillot, S., Negredo, A.M., Villaseñor, A. (2014). The coupling of Indian subduction and Asian continental tectonics, Invited Focus Review. *Gondwana Research*. doi: 10.1016/j.gr.2014.04.003.

Replumaz, A., Funiciello, F., Reitano, R., Faccenna, C., Balon, M. (2016). Double subduction of continental lithosphere, a key to form wide plateau. *Geology*. doi: 10.1130/G38276.1.

Schneider, F.M., Yuan, X., Schurr, B., Mechie, J., Sippl, C., Haberland, C., Minaev, V., Oimahmadov, I., Gadoev, M., Radjabov, N. et al. (2013). Seismic imaging of subducting continental lower crust beneath the Pamir. *Earth and Planetary Science Letters*, 375, 101–112. doi: 10.1016/j.epsl.2013.05.015.

Sobel, E.R., Chen, J., Schoenbohm, L.M., Thiede, R., Stockli, D.F., Sudo, M., Strecker, M.R. (2013). Oceanic-style subduction controls late Cenozoic deformation of the Northern Pamir orogen. *Earth and Planetary Science Letters*, 363, 204–218. doi: 10.1016/j.epsl.2012.12.009.

Spurlin, M.S., Yin, A., Horton, B.K., Zhou, J., Wang, J. (2005). Structural evolution of the Yushu-Nangqian region and its relationship to syncollisional igneous activity, east-central Tibet. *Geological Society of America Bulletin*, 117, 1293–1317.

Tapponnier, P. and Molnar, P. (1976). Slip line field theory and large-scale continental tectonics. *Nature*, 264, 319–324. doi: 10.1038/264319a0.

Vanderhaeghe, O. (2012). The thermal–mechanical evolution of crustal orogenic belts at convergent plate boundaries: A reappraisal of the orogenic cycle. *Journal of Geodynamics*, 56–57, 124–145. doi: 10.1016/j.jog.2011.10.004.

van der Voo, R., Spakman, W., Bijwaard, H. (1999). Tethyan subducted slabs under India. *Earth and Planetary Science Letters*, 171, 7–20.

Wang, M. and Shen, Z.-K. (2020). Present-day crustal deformation of continental China derived from GPS and its tectonic implications. *Journal of Geophysical Research*, 125. doi: 10.1029/2019JB018774.

Wang, S., Replumaz, A., Chevalier, M.-L., Li, H. (2021). Decoupling between upper crustal deformation of southern Tibet and underthrusting of Indian lithosphere. *Terra Nova*, 1–10. doi: 10.1111/ter.12563.

Wong, A., Ton, S.Y.M., Wortel, M.J.R. (1997). Slab detachment in continental collision zones: An analysis of controlling parameters. *Geophysical Research Letters*, 24(16), 2095–2098.

Zhao, W., Kumar, P., Mechie, J., Kind, R., Meissner, R., Wu, Z., Shi, D., Su, H., Xue, G., Karplus, M. (2011). Tibetan plate overriding the Asian plate in central and northern Tibet. *Nature Geoscience*, 4, 870–873.

# 3

# The Major Thrust Faults and Shear Zones

Djordje GRUJIC and Isabelle COUTAND

*Dalhousie University, Halifax, Canada*

## 3.1. Introduction

The Himalayan range is underlain by a basal detachment which is exposed at the surface along the southern foothills, plunges northwards for at least a couple hundreds of kilometers and extends laterally for more than 2,000 km along the orogen's length. Because of its size and capacity to generate earthquakes up to magnitude 9, the Himalayan basal detachment is a megathrust akin to the megathrusts in subduction zones. In orogens, contractional deformation propagates from the core of the mountain belt towards its peripheral foreland. Such in-sequence deformation is manifested by the successive activation of thrust faults and shear zones from the hinterland towards the foreland (Boyer and Elliott 1982; Dahlstrom 1970). In this respect, the Himalaya are similar to other orogenic belts. The main shear zones and faults currently exposed at the surface were formed over the last 30 million years, from the oldest structure in the North, in the interior of the orogen, towards the currently active fault exposed along the southern foothills and carrying the entire orogen over the Indus–Ganges–Brahmaputra

*Himalaya, Dynamics of a Giant 1*,
coordinated by Rodolphe CATTIN and Jean-Luc EPARD.
© ISTE Ltd 2023.

alluvial planes (see Volume 2). These structures were active at different depths under different pressure and temperature conditions, ranging from granulite metamorphic grade to near-surface conditions. Accurate interpretations of these structures therefore entail the knowledge of their deformation conditions. Consequently, the Himalaya offer a prime natural laboratory to investigate the development and the interaction of structures operating at different crustal levels. Of particular societal interest is the understanding of the interactions between deep aseismic movements and near-surface seismic slip generating earthquakes.

## 3.2. Some basic concepts

In this section, we define key structural concepts and terms relevant to continental deformation discussed in this chapter.

Faults versus shear zones: in the upper crust, at deformation conditions below $\sim 350°C$[1], the brittle failure and reactivation of pre-existing planes are controlled by differential stress, cohesion, pore fluid pressure and friction. We call these structures faults, and the rocks affected by faulting are cataclasites. At deeper crustal levels, deformation is dominated by thermally activated crystal-plastic deformation mechanisms. Structures formed under such deformation conditions are shear zones, and the resulting deformed rocks are mylonites. Faults are either single or multiple planes (fault zones) and zones of cataclastically deformed rocks (damage zones) that range from a few centimeters to hundreds of meters thick. Shear zones are sometimes only a few centimeters thick, but can reach several kilometers in thickness and accommodate hundreds of kilometers of displacement without the loss of rock cohesion. Because brittle and crystal-plastic structures respond differently to stresses, using a unique structural or tectonic model to interpret an orogen like the Himalaya misrepresents its structural complexity.

Detachment: low-angle or horizontal fault or shear zone separating an upper plate (hanging wall) from a lower plate (footwall). Décollement is either a synonym, or a large-scale detachment. The terms are used in both extensional and contractional tectonic settings. Both terms are also used to describe the sole thrust that separates the stack of thrust nappes from the less deformed or undeformed basement. However, note that faults and shear zones with dip-slip

---

1. The temperature of the brittle–ductile transition depends on a given mineral, its water content, fault kinematics, strain rate, pore fluid pressure.

movement are neither extensional nor contractional; the kinematics along such structures is best indicated with top-to-the-geographic direction.

Elements of a shear zone: outside the shear zone boundaries, there is no perceivable deformation. Inside the shear zone, the finite strain increases towards the core of the shear zone and the axes of the finite strain ellipsoid rotate towards parallelism with the shear zone boundaries. Various lithologies can be found within a single crustal scale shear zone, with mylonitic textures in the core of the shear zone to proto-mylonites textures towards the margins where finite strain vanishes. Therefore, the rocks within a shear zones should be described in terms of their protoliths, and the boundaries of mylonites derived from different lithologies are protolith boundaries. The core of a shear zone is the structure that is usually indicated on the geologic maps but the shear zone boundaries are seldomly mapped. These circumstances cause misunderstandings and disagreements among geologists.

Age of tectonic structures: geological structures are (or have been) active over a period of time, from thousands to tens of millions of years. Therefore, the structure's timing of activity, that is, when they started and ceased to operate, covers a time range and not a single date as is often published.

Rock uplift, surface uplift and exhumation: we follow the definitions outlined by England and Molnar (1990).

## 3.3. Main faults and shear zones

The Himalayan orogenesis started during the subduction of the Tethys ocean underneath the Eurasian plate forming an Andean-type continental margin. Remnants of the subduction zone are preserved in an ophiolitic belt (Yarlung–Tsangpo suture zone) and associated accretionary wedge and forearc basin sediments (Xigazê/Shigatse basin sediments, and Dras–Nindam unit or Zanskar flysch). Mountain building propagated southwards, in-sequence, first affecting the distal sediments deposited along the northern passive margin of the Indian plate, the Tethyan Sedimentary Sequence, and progressively involved sediments increasingly proximal to the Indian Plate. The structures generated by these early stages will not be discussed in detail in this paper. As shown in Figure 3.1 the Himalaya sensu stricto are defined by the four first-order tectonic boundaries (Gansser 1964): South Tibetan detachment system (STDS), Main Central thrust (MCT), Main Boundary thrust (MBT) and the Main Frontal thrust (MFT). Each of them is, in general, shallowly to gently north dipping and has unique characteristics due to the orogen's size, internal temperature, slip rate, and also the asymmetric distribution and rate of surface denudation.

### 3.3.1. *South Tibetan detachment system (STDS)*

The South Tibetan detachment system (STDS) consists of one or two parallel shallow north dipping ductile shear zones cut by steeper brittle normal faults. The dominant kinematics (sense of shear) is top-down-to-the north; therefore, the STDS has a normal-sense kinematics. The ductile shear zone juxtaposes the high-grade metamorphic rocks of the Greater Himalayan sequence (GHS) in its footwall (see Volume 2 – Chapter 5), and the low-grade metasediments of the Tethyan sedimentary sequence (TSS) in its hanging wall (see Volume 2 – Chapter 3). STDS runs nearly across the highest Himalayan peaks (Figure 3.1) from 75° E to 94° E (or over a length of nearly 2,000 km). In the westernmost segment of the orogen, in the Zanskar region of Ladakh (see Volume 2 – Chapter 2), it is named the Zanskar shear zone (Epard and Steck 2004). In the Sutlej section, the Sangla detachment correlates with the STDS. Between the two regions, for about 100 km along the strike, the presence and location of the STDS are disputed (Stübner et al. 2014), and the low-grade Haimantas Fm. (Epard and Steck 2004) are attributed either to the GHS (e.g. Steck 2003) or to the TSS (e.g. Webb et al. 2011). Therefore, the STDS either does not exist (or did not reach the surface), or the rock from its hanging wall block extends until the foothills, and completely covers the GHS.

Because the hanging wall block of the STDS consists of a continuous, right way up stratigraphic sequence resting on top of metamorphic rocks, geologists did not realize the existence of the major shear zone but assumed it was a "gradual" nonconformity (Heim and Gansser 1939; Gansser 1983). Until the early 1980s, ductile shear zones and mylonites were rather unknown (Ramsay 1980). In addition, before the first geochronological dates became available (Hamet and Allègre 1976; Ferrara et al. 1983), the paragneisses of the GHS were assumed to be older (Archaean, Heim and Gansser 1939), than the TSS. Finally, the logistical and political inaccessibility of the border areas between Tibet and the Himalayan states did not allow extensive and systematic field work in the vicinity of the STDS. The STDS was identified as a shear zone only in the early 1980s (Burg et al. 1984; Brun et al. 1985; Burchfiel and Royden 1985; Herren 1987; Kündig 1988; Pêcher 1991).

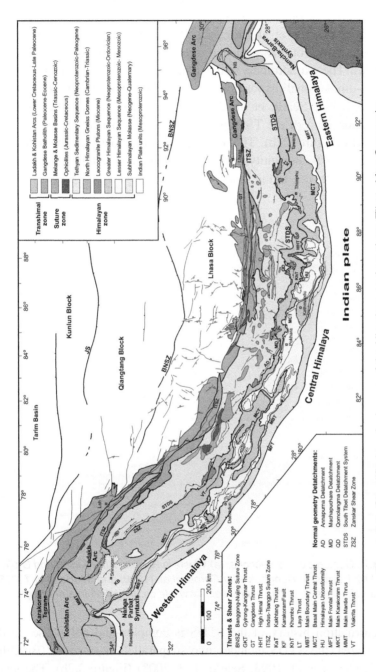

**Figure 3.1.** *Simplified geologic map of the Himalaya and southern Tibet. After Goscombe et al. (2006). For a color version of this figure, see www.iste.co.uk/cattin/himalaya1.zip*

This STDS was eventually folded into gentle upright folds, and subsequent erosion has left erosional remnants, klippen, of the TSS rocks on top of the GHS rocks (Kellett et al. 2009; La Roche et al. 2018) preserved tens of kilometers south of the principal modern surface trace of the STDS (Figure 3.2a). The extraordinary relief around the highest Himalayan peaks offers three-dimensional exposures of the STDS (Burchfiel et al. 1992; Carosi et al. 1998; Searle et al. 2003) which suggest that the extent of tectonic juxtaposition of the TSS against the GHS is on the order of at least 200 kilometers (Figure 3.2b). This overlap does not constrain the amount of slip along the STDS, a quantity that is impossible to determine in the case of normal detachments due to the absence of markers.

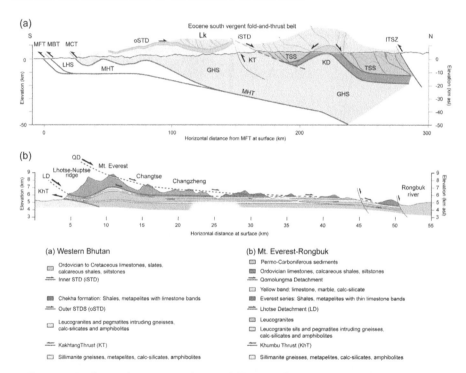

**Figure 3.2.** *Geologic cross-sections. (a) Western Bhutan. After Kellett and Grujic (2012). (b) Mt Everest to Rongbuk after Carosi et al. (1998) and Searle et al. (2003). For a color version of this figure, see www.iste.co.uk/cattin/himalaya1.zip*

The deformation history of the STDS is complex and polyphase. In the Mt Everest area, the STDS includes two subparallel shear zones (Burchfiel et al. 1992; Carosi et al. 1998; Searle et al. 2003) (Figure 3.2b). The lower one, the Lhotse detachment, separates the high-grade metamorphic rocks (gneisses, migmatites, calc-silicates) and granitoids of the GHS from the TSS sediments in the hanging wall. The upper one, the Qomolungma[2] detachment (Figure 3.3), is located within the TSS. The Lhotse detachment is a couple of kilometer thick ductile shear zone (Carosi et al. 1998; Cottle et al. 2011), while the structurally higher Qomolangma detachment is dominantly brittle (Carosi et al. 1998; Cottle et al. 2011). This change in deformation mechanisms may result from a decrease of the peak temperatures across the ductile shear zone, from about 650°C at the base of the Lhotse detachment to about 475°C at the top of it (Cottle et al. 2011; Law et al. 2011; Waters et al. 2019), and to about 350°C in the hanging wall of the Qomolangma detachment (Cottle et al. 2011). A similar temperature range across the STDS was observed elsewhere in the Himalaya (Kellett and Grujic 2012; Parsons et al. 2016; Soucy La Roche et al. 2018; Long et al. 2019). The combined thermobarometry data indicate an apparent temperature gradient of about 375°C/km (Kellett et al. 2019) across the ductile shear zone of the STDS, which requires an interplay between the perturbation of the temperature field due to the advection of heat in the footwall block (e.g. Campani et al. 2010), and shearing of deformed isotherms (Law et al. 2011). The advection of heat by tectonic rock uplift (in general or in case of a detachment's footwall) would condense isotherms near the surface (Braun 2016), that is, at the top of the shear zone, while the ductile shearing would bring them even closer causing apparent temperature gradients exceeding 400°C/km (Law et al. 2011). Because of the lower temperature of deformation, the upper Qomolungma detachment is usually interpreted as being younger than the structurally lower Lhotse ductile detachment. However, because of the steep temperature gradient across the STDS, it is possible that the two detachments operated simultaneously.

During the Miocene, the GHS rocks have undergone amphibolite to granulite facies metamorphism, anatexis and magmatism that have strongly overprinted and even erased the earliest Himalayan microstructures (see Volume 2 – Chapter 6). Nevertheless, structural investigations of Miocene leucogranites affected by the STDS have revealed two superposed sets of north-dipping structures; an earlier set indicating a top to the south sense

---

2. Qomolungma is the Tibetan name for the Mt. Everest, also spelled Chomolangma, meaning "Goddess Mother of the World". The Nepali name for Mt. Everest is Sagarmatha, meaning "Goddess of the Sky".

of shear (i.e. thrusting) overprinted by a subsequent normal sense of shear corresponding to the South Tibetan Detachment's kinematics (Finch et al. 2014). In the shear zone itself, there was no anatexis, and magmatism is restricted to local injections of Miocene leucogranite dykes and sills (Figure 3.4a). In contrast, within the TSS, there is a pervasive earlier deformation caused by the deformation of the TSS in a south vergent fold-and-thrust belt (Figure 3.2a, Ratschbacher et al. 1994; Wiesmayr and Grasemann 2002; Kellett and Grujic 2012; Finch et al. 2014). Although there is little field evidence, we can reasonably suggest that the TSS fold-and-thrust belt was soled by a south-vergent basal detachment separating passive margin sediments of the TSS from the basement of the Indian plate (Wiesmayr and Grasemann 2002). The base of the detachment was also associated with the development of an SW-verging Fang nappe in the Annapurna massif (Colchen et al. 1986; Vannay and Hodges 1996). That detachment was reactivated during the Miocene with an opposite sense kinematics to create the STDS (Kellett and Grujic 2012; Finch et al. 2014).

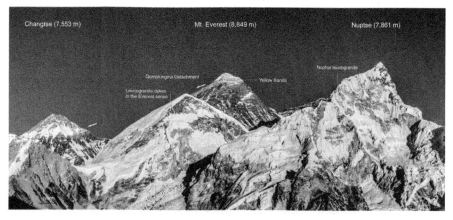

**Figure 3.3.** *Qomolangma detachment. Notice the Yellow bands and the darker Everest series beneath it. The Lhotse Detachment is hidden by the relief but it is located between Everest series and the Nuptse granite. Observe several generations of slightly deformed dykes in the roof of the Nuptse leucogranite. Nuptse-Mt.Everest-Changtse panorama by Daniel Hug. For a color version of this figure, see www.iste.co.uk/cattin/himalaya1.zip*

Thrusting and normal shearing were coplanar and codirectional, with S-directed thrusting overprinted by N-directed normal shearing (SW and NE in the western Himalaya). This inversion of sense of movement occurred between

26 and 20 Ma (Finch et al. 2014). Several authors have observed opposite dipping shear bands, suggesting the switch of top to the south to top to the north shearing. However, the published micrographs do not indicate crosscutting relationships (Figure 3.4f), and it is possible that the two sets of shear bands are actually conjugate (Grujic et al. 2002), and result from a transpression (simple shear combined with sub-vertical shortening) within the STDS.

After the passage through the ductile–brittle transition, the STDS mylonites were overprinted by cataclasites. To the best of our knowledge, there is no study on this transition in the deformation mechanisms along the STDS, but an excellent example has been described in detail in the central Alps of Switzerland along the Simplon fault (Mancktelow 1990). There, as the footwall block of the shear zone was progressively exhuming, the mylonites cooled below the brittle–ductile transition and were overprinted by cataclasites. At locations where two parallel detachments were mapped like at Mt Everest, the upper brittle one was interpreted as younger, although the compilation of geochronological data does not exclude the possibility that the two detachments were coeval, at least temporarily. Finally, the STDS shear zone was subsequently cut by steep normal faults (Figure 3.2b). Some of the brittle faults cross-cut Quaternary glacial sediments and may be seismically active (Hurtado et al. 2001; Meyer et al. 2006; Rana et al. 2013). It has not been determined whether these faults are caused by the same tectonic processes that created the STDS.

The age range of the STDS activity was determined by geochronology and thermochronology on the leucogranites affected by or cutting this structure. The oldest crystallization ages are approximately 26 Ma and the youngest approximately 10 Ma (Figure 3.5) (Kellett et al. 2009, 2013; Rubatto et al. 2012). Low-temperature multisystem thermochronology indicates the cessation of ductile shearing at 10–11 Ma (Kellett et al. 2013; Shurtleff 2015). According to the data available, there is an apparent eastward younging of the cessation of ductile motion along the STDS (Leloup et al. 2010). Considering the complexities of the deformation history and strain partitioning (two closely spaced sub-parallel shear zones, ductile and brittle deformation, coeval or in succession), the available age data may not yet be yielding the accurate deformation history of the STDS as further systematic work is needed.

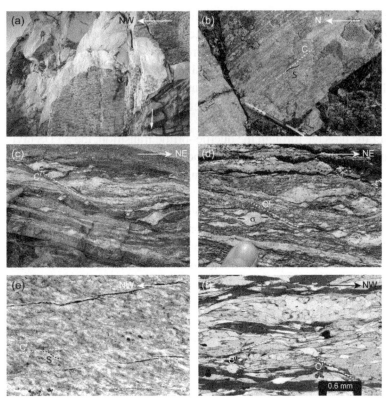

**Figure 3.4.** *Structures from the South Tibetan detachment system. (a) Lhotse detachment. Top: the GHS beneath the South Col. Several generations of cross cutting dykes of variable composition. Earliest dykes in the calc-silicates (lower right bottom with arrow) are boudins. Climber is Jost Kobusch, photo by Daniel Hug (2020). (b) South Tibetan detachment. Mylonitic leucogranite. Dominant planar fabric are the N dipping C planes. North Sikkim, photo by Djordje Grujic. (c) Zanskar Shear Zone. S-C fabric and delta clasts from breaking up of leucogranite dykes, cut by a synthetic shear zone parallel to C'. Photo by Roberto Weinberg. (d) Zanskar Shear Zone. C' in porphyritic orthogneiss. Narrow, sheared leucogranite dykes (at the finger). Reru, Zanskar. Photo by Roberto Weinberg. (e) South Tibetan Detachment. Sheared tourmaline and garnet bearing leucogranite. C planes dipping to the left. NW Bhutan. Photo by Djordje Grujic. (f) Photomicrograph of a two mica, garnet, staurolite schist from the Chekha formation. Note two sets of opposite-dipping conjugate shear bands, one indicating top-to-the south, the other top-to the north shearing. This geometry of microstructures indicates a component of pure shear (shortening perpendicular to foliation), rather than two successive kinematic events. For a color version of this figure, see www.iste.co.uk/cattin/himalaya1.zip*

### 3.3.2. *Main Central thrust (MCT)*

The MCT is the most epitomic Himalayan structure. In general, the first observation of thrusting in the Himalaya was made by Mallet (1875). The original definition of the MCT was structural and metamorphic; the MCT was firstly defined as a thrust that produced a marked break in metamorphic grade between higher-grade hanging wall and lower-grade footwall rocks (Heim and Gansser 1939).

It has been long recognized that the MCT is not a single plane and not a fault. The changes in deformation style were found to be gradual (Pêcher 1977). Bordet (1961) noted lithological repetitions and grouped them into his "zone des écailles", and Arita et al. (1973) conceived the "main central thrust zone" consisting of three thrust sheets. Already, Le Fort (1975) inferred the thrust mechanism to be a plastic deformation by "intracrystalline gliding"[3], distributed through a broad zone, comparable to gliding in a pack of cards and consistent with the main schistosity and the mineral "streak lineation"[4] in the direction of transport. Unfortunately, the incorrect "fault" terminology persists in the modern literature about the MCT, biasing the interpretation of the tectonics of the Himalaya.

The MCT is a north-dipping shear zone with a top to the South sense of displacement (thrust motion), separating the GHS rocks in the hanging wall (see Volume 2 – Chapter 5) from LHS rocks in the footwall block (Figure 3.1 and Volume 2 – Chapter 7). Both rock sequences are from the Indian plate, with the time overlap between Paleoproterozoic and Paleozoic Lesser Himalayan protolith deposition proximal to India and Greater Himalayan protolith and Tethyan Himalayan sediment deposition distal to India (Long et al. 2011c). The magmatic and volcanic rocks in the LHS are Paleoproterozoic and most likely represent a 1,830 $\pm$ 50 Ma continental arc (Kohn et al. 2010). The orthogneisses in the GHS are Neoproterozoic and Cambro-Ordovician (Cawood et al. 2007). The presence of Paleoproterozoic orthogneisses in the

---

3. Development of crystallographic preferred orientation by intracrystalline processes Passchier, C.W. and Trouw, R.A. (2006). *Microtectonics*, 2nd edition. Springer-Verlag, Berlin, Heidelberg, New York.
4. The authors were likely referring to stretching and perhaps mineral lineation.

GHS is not excluded as zircon xenocrysts of that age have been observed in few studies (Cottle et al. 2009; Chakungal et al. 2010). Bulk rock geochemistry has also been used to define the location of the MCT (Godin et al. 2021), or in the terminology that we use in this chapter, the protolith boundary between the GHS and LHS rocks.

There are no cut-off lines[5]; therefore, it is difficult to constrain the displacement along the MCT. Heim and Gansser (1939) estimated that, in the Kumaon Himalaya, it is on the order of 300 km based on the superposition of the GHS over the LHS rocks. Studies using balanced cross-sectional restoration, and therefore assuming the GHS is a rigid thrust sheet and the MCT is a single plane fault, yielded estimates of horizontal offset ranging from ∼90 to 400 km (Schelling and Arita 1991; Long et al. 2011a; Webb 2013; Long et al. 2012, 2016; Robinson and Martin 2014). According to geodynamic models, the offset across the MCT may be on the order of 400 km or more (Jamieson et al. 2006). Petrochronology and thermochronology data indicate that the MCT started to operate at the Oligocene–Miocene boundary (∼23 Ma) and ceased its activity at 11 Ma (Figure 3.5). Therefore, at the scale of the orogen, displacements on the STDS and the MCT were coeval during the Early to Middle Miocene (Figure 3.5).

The MCT places amphibolite to granulite metamorphic grade GHS rocks on top of greenschist to lower metamorphic grade LHS rocks, which is the basic characteristic of a thrust. Contrary to the original metamorphic definition of the MCT, modern thermobarometric data indicate that there is no jump in the peak metamorphic grade across the MCT (Le Fort 1975; Pêcher 1977, 1978; Brunel 1986; Goscombe et al. 2018; Grujic et al. 2020) as it might be observed across a thrust fault. Instead, peak metamorphic temperatures continuously increase across the entire LHS (from bottom to top) and the lower part of the GHS, in the form of an inverted metamorphic (or temperature) field. Peak metamorphic temperatures increase from about 450°C a couple of kilometers structurally below the MCT to > 650°C at the MCT or less than a kilometer above it.

---

5. A cut-off line is the intersection line between a fault surface and the reference marker in each fault block. The distance between the cut-off lines in the slip direction is the fault slip magnitude.

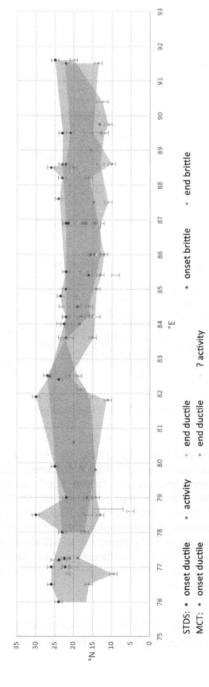

**Figure 3.5.** *Compilation of age data constraining the timing of activity of the STDS (blue-shaded area) and the MCT (red-shaded area) along the strike of the Himalaya. For a color version of this figure, see www.iste.co.uk/cattin/himalaya1.zip*

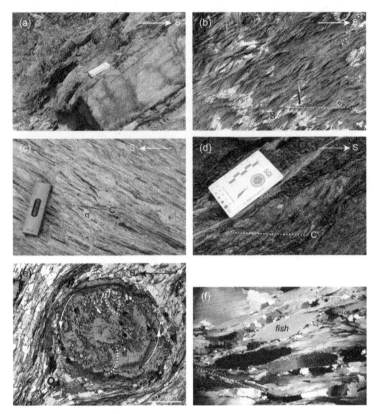

**Figure 3.6.** *Structures from the main central thrust zone. (a) Sharp contact between mylonitized Cambro-Ordovician granitoids in the hanging wall and mylonitic Paleoproterozoic quartzites in the footwall. Eastern Bhutan. Photo by Djordje Grujic. (b) Coarse S-C fabric in the schist of the LHS. Subhorizontal shear bands (C' planes) indicate top to the SW shearing. Photo by Roberto Weinberg. (c) Mylonitic Proterozoic augengneiss within the LHS. Asymmetric feldspar porphyroclasts and weak subhorizontal C' planes indicate top to the South shear. Photo from GeoSciTweeps (Twitter) (d) S-C fabric in the garnet schist of the LHS. Slight subhorizontal shear bands (C' planes) indicate top to the S shearing. Photo by Salvatore Iaccarino. (e) Plane-polarized light photomicrograph of rotated inclusion trails in a snowball garnet porphyroblast. Asymmetry shows top-to-SW shear sense (Roberts et al. 2020). (f) Gently south dipping shear bands define mica fishes and indicate top-to-the south shearing. Photo by Chiara Montomoli. For a color version of this figure, see www.iste.co.uk/cattin/himalaya1.zip*

Deformation at these temperatures leads to mylonitization of both LHS and GHS rocks. Because of decreasing temperatures down the section, the

LHS deformation becomes more localized, yet microstructural observations demonstrate that deformation was achieved by dynamic recrystallization (mylonitization) across the topmost 2–4 km of the LHS (Brunel 1986; Long et al. 2011b, 2016; Larson et al. 2017; Grujic et al. 2020). This deformation affected Proterozoic quartzites, schists and granitoid gneisses (Figure 3.6). Because of both changing lithologies and the progressive localization of deformation, it is difficult to map the lower boundary of the MCT shear zone. Conversely, in the hanging wall, the peak metamorphic temperature increases up the section across the entire heterogeneously sheared GHS unit, which also prevents mapping the shear zone's upper boundary (Law et al. 2013). Nevertheless, the MCT-related a high strain zone, that is, the mylonitic belt in the hanging wall, is at least one kilometer thick. The "main" central thrust becomes an arbitrary median plane in a broad zone of crystal-plastic deformation (Figure 3.7) and progressive metamorphic changes (Le Fort 1975; Stöcklin 1980). The MCT is therefore a several km-thick mylonitized zone affecting different lithologies both in the hanging wall and footwall blocks.

Like in a typical shear zone, the finite strain increases towards the core of the shear zone, but this trend is modified by the type of shear zone growth (Fossen and Cavalcante 2017), and by strain localizations at lithological boundaries. Within the mylonite belt, there are several protolith boundaries which were given regional names. The most important of them separates GHS rocks (Neoproterozoic and Paleozoic magmatic and sedimentary protoliths) from LHS rocks (Paleoproterozoic sedimentary and magmatic and volcanic rocks). This is the boundary (Figure 3.6a) most field geologists try to identify in the field and map as the MCT. However, the protoliths astride the shear zone may be very similar (e.g. schist against schist). In such locations, some authors use the kyanite-in isograd as the proxy for the MCT position, since in locations where the MCT is well defined (e.g. Cambro-Ordovician granitic gneiss on top of Paleoproterozoic metasedimentary rocks), the kyanite isograd is parallel to it.

In summary, to appropriately represent the MCT, we recommend mapping both the shear zone and the protolith boundaries in it (e.g. Hunter et al. 2018). An excellent example is the map of the Insubric Line (i.e. shear zone) (Schmid et al. 1987) in the central Alps.

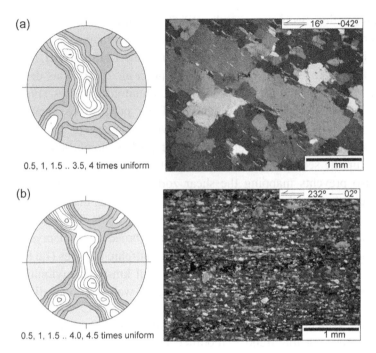

**Figure 3.7.** *Optically measured quartz c-axis fabrics and recrystallization microstructures in quartz-rich mylonitic gneiss of the GHS from the Sutlej valley. Fabric and microstructural analyses are used to determine the sense of shear, geometry of deformation and deformation conditions (temperature and flow stresses). All the published data of this type in the Himalaya indicate that crystal plastic deformation occurred for at least 4 km below the MCT, across the entire GHS and at least 1 km above the ductile STDS. (a) GHS quartz-rich paragneisses at ~20 m above the MCT. Deformation temperature is ~610°C. (b) GHS quartz-rich mylonitic orthogneiss at ~70 m above the MCT. Deformation temperature is ~535°C. The temperature difference at the same structural level because of the more internal position of outcrop (a). Fabric asymmetries and microstructures show a top to the SW (thrust) shear sense. From Law et al. (2013). All fabric diagrams and thin sections oriented perpendicular to foliation, parallel to lineation and viewed towards the NW; equal-area lower-hemisphere projections. For a color version of this figure, see www.iste.co.uk/cattin/himalaya1.zip*

### 3.3.2.1. *High Himalayan discontinuity*

The pervasive ductile shear of the GHS is heterogeneously distributed, and there are localized levels of higher strain, that is, internal shear zones in it. Only a few of them have significant displacement manifested by a jump in metamorphic grade and the age of Himalayan magmatic rocks. Some of them are out-of-sequence, the slip started later than the slip along the MCT and lasted after the slip along the MCT ended. In contrast, some of them appear to have formed in sequence with the MCT. These shear zones are apparently not as continuous and extensive as the STDS and the MCT; they consist of segments that likely were asynchronously active. The amount of displacement was much less than that along the MCT, and in most cases field structural evidence is lacking but there is a jump in metamorphic grade (Ganguly et al. 2000; Grujic et al. 2011) and in the age of leucogranites across these cryptic structures (Rubatto et al. 2012); hence, these structures are often named discontinuities.

A prime example of these out-of-sequence shear zones within the GHS is the Kakhtang thrust in Bhutan (Figure 3.8) (Gansser 1983). The thrust was active at 14–10 Ma (Grujic et al. 2002), concurrently with the slip of both STDS and MCT. It placed granulite facies metamorphic rocks with remnants of mafic eclogites on top of amphibolite facies rocks (Swapp and Hollister 1989; Grujic et al. 2011; Warren et al. 2011), and there is a jump in both metamorphic and magmatic ages across it; in the hanging wall, the magmatic ages range from 11 to 15 Ma, while in the footwall, they range from 16 to 25 Ma (Carosi et al. 2006; Grujic et al. 2011; Warren et al. 2011; Montomoli et al. 2017). A similar intra-GHS structure exists in northern Sikkim where we can observe a similar jump in magmatic ages and metamorphic grade (Ganguly et al. 2000; Rubatto et al. 2012; Kellett et al. 2013). It was named the Zema Discontinuity (Chakraborty et al. 2016), but was not yet observed in the field. Other out-of-sequence thrusts within the GHS include the Kalopani shear zone (Vannay and Hodges 1996) in the Annapurna range and the Khumbu thrust in the Mt. Everest massif (Searle 1999). The latter is not the same structure as the Khumbu thrust by (Valdiya 1980).

In eastern Nepal, a thrust was mapped within the GHS by Goscombe et al. (2006) and named Higher Himal Thrust. An equivalent "tectono-metamorphic" discontinuity was recognized in western Nepal, to the west of the Annapurna–Dhaulagiri range, and named high Himalayan discontinuity (HHD) (Montomoli et al., 2013). The structure is rarely observed in the field but there is a clear change of metamorphic conditions astride it, and indications of an in-sequence deformation relative to the MCT. The structure was mapped across most of western Nepal (Carosi et al. 2018), until Kumaun Himalaya in India (Benetti et al. 2021).

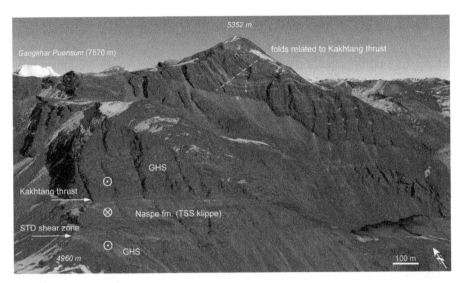

**Figure 3.8.** *Kakhtang thrust in north-central Bhutan. This and equivalent structures are difficult to identify in the field because of the similar rocks in their footwall and hanging walls. At this locality, the footwall contains a sliver of graphitic garnet–staurolite schists, the Naspe formation (Gansser 1983), which belongs to a klippe of Tethyan sedimentary sequence (Kellett et al. 2009). The base of the klippe is the South Tibetan detachment that separates The Naspe formation from sillimanite bearing gneisses and migmatites of the GHS formation in the footwall. Image from Google Earth. Scale of the foreground is indicated by the scale bar. For a color version of this figure, see www.iste.co.uk/cattin/himalaya1.zip*

### 3.3.3. *Main Boundary thrust (MBT)*

The northern limit of deposition of the Siwaliks group, the Neogene Himalayan foreland sediments, was identified and named by H. Medlicott (1859, in Oldham 1893) the "main boundary fault", as the northern "boundary" of the sedimentary basin. Following the field observations by Mallet (1875), and their own, Heim and Gansser (1939) established great horizontal displacement along this fault in the Kumaon Himalaya and the Darjeeling Himalaya, and named it the main boundary fault, or main boundary thrust (MBT).

The MBT is a brittle to brittle–ductile fault that places the Paleozoic LHS rocks on top of the syn-orogenic Himalayan foreland sediments, mostly the Siwaliks group (Figure 3.1 and Volume 2 – Chapters 7 and 8). Where exposed, the MBT is generally marked by a narrow (∼100 m or less) zone of cataclasis that typically moderately to steeply dips northward (Figure 3.9). Locally, Siwalik sediments may be missing and the LHS rocks are thrust directly over Quaternary alluvial sediments and perhaps even over the basement of the Indian plate (Heim and Gansser 1939). In the vicinity of the MBT trace, several other thrusts have been mapped (e.g. Thakur et al. 2010), but in this chapter, we only refer to the MBT sensu stricto. This thrust has yet to be directly dated; indirect thermochronological studies suggest that thrusting started after 10 Ma (Meigs et al. 1995), or at 13–14 Ma (Singh and Patel 2022). The former date is consistent with the time of cessation of slip along the MCT and the propagation of horizontal shortening to the MBT. Assuming the Himalayan orogen was entirely governed by Coulomb wedge tectonics, sequential, balanced cross-sectional restoration aided by thermochronological data suggests that the MBT initiated between 10 and 8 Ma (Long et al. 2012; Robinson and McQuarrie 2012; McQuarrie et al. 2019). Equivalent structural analyses of the LHS also suggest that the fold-and-thrust deformation style developed in sequence and that the MBT is just the last of the intra-LHS thrusts (Long et al. 2012; McQuarrie et al. 2019). Estimates of cumulative shortening across the LHS range from 200 to 400 km, and the displacement along the MBT is not significantly larger than that along most of the other thrusts within the LHS.

The MBT is the surface expression of a frontal ramp, and it flattens out at about 10–12 km depth until a deeper ramp located about 100 km north from the orogenic front occurs. Deformation along the flat segment of this detachment is seismic (Avouac 2015); hence, the décollement and the wedge above it deform in a brittle–frictional manner. According to this observation and the presence of a duplex in the outer LHS, the MBT is interpreted as being the frontal thrust of the Late Miocene–Pliocene Coulomb wedge of the Himalaya.

The slip along the MBT apparently ceased when the deformation propagated to the main frontal thrust, about 2 Myr (Mugnier et al. 2004; van der Beek et al. 2006) or 4–5 Myr (McQuarrie et al. 2019) ago. However, there have been recent movements close to the boundary between the Siwaliks and the outer LHS: normal and strike-slip faulting as well as thrust faulting (Nakata 1989; Yeats and Lillie 1991; Mugnier et al. 1994; Thiede et al. 2017). Variations in several parameters (decrease in the basal slope or increase in the topographical slope or of the pore fluid pressure) could have transformed a stable Himalayan Coulomb wedge into an over-critical one (Mugnier et al. 1994) causing internal deformation of the wedge. Alternatively, the

MBT-parallel normal faults in the outer LHS were interpreted as extensional deformation within the hanging wall of a megathrust system following a major earthquake (Riesner et al. 2021).

**Figure 3.9.** *Main boundary thrust. (a) Gondwana coal-bearing clastic sandstone from the Lesser Himalayan Sequence in the hanging wall; conglomerates of the upper Siwalik (Pleistocene) in the footwall. Fault is marked by a 5–30 cm thick layer cataclasites and fault gouge derived from hanging wall rocks. (b) Same fault seen on a horizontal surface. Coal seam at the top separated from conglomerates at the bottom by a wedge of foliated cataclasite. Rishore coal mine, east Bhutan. Photos by Djordje Grujic. For a color version of this figure, see www.iste.co.uk/cattin/himalaya1.zip*

### 3.3.3.1. *Munsiari–Ramgarh–Shumar thrust system*

It has long been known that the LHS bounded by the MCT above and the MBT at the base is a structurally complex tectonic unit. The structurally upper part contains Paleoproterozoic metasediments, magmatic and volcanic rocks, and has undergone greenschist grade metamorphism. This segment of the LHS has been named lower (stratigraphically) or inner (within the orogen) LHS. The structurally lower unit preserves the original bedding in dominantly Neoproterozoic to late Paleozoic sedimentary rocks permitting the mapping of numerous horses stacked in different duplex styles. It was named upper (stratigraphically) or outer (within the orogen) LHS. There is about 1 billion years hiatus between the upper and lower LHS groups (Long et al. 2011c). The boundary between the two is from West to East, the Munsiari, Ramgarh and Shumar thrusts (MRST) in western India, Nepal and Sikkim, and Bhutan, respectively, which may be a continuous structure at the scale of the orogen (Pearson and DeCelles 2005; Robinson and Pearson 2013). The timing of the Munsiari, Ramgarh and Shumar thrusts and their role in Himalayan tectonics are disputed (Stübner et al. 2014). Although most geologists agree on the existence of duplexes in the LHS, there is no agreement on their time of formation, on the significance of the MRST thrusts or on the relative timing of deformation (i.e. the sequence of MCT–MRST–MBT). The MRST is a passive roof thrust[6] of the LHS duplex and has accommodated on the order of 100–120 km of horizontal shortening, comparable to the shortening across the MCT and possibly more than that across the MBT (Pearson and DeCelles 2005; Long et al. 2011a; Robinson and McQuarrie 2012; Mandal et al. 2019). Other studies equate the MRST with the MCT shear zone (Searle et al. 2008; Larson and Godin 2009). According to our discussion about the strain in the MCT zone, the pervasive top-to-the-south ductile shearing and the inverted temperature gradient extends down until the MRST (at least in the eastern Himalaya), but the thrust is not the MCT shear zone boundary, as it is a younger structure.

Estimates of the timing of activity of the MRST vary along the orogen. Based on structural reconstructions, early Miocene ages ($\sim$20–15 Ma) have been suggested for western Nepal and eastern Bhutan (Long et al. 2011a; Robinson and McQuarrie 2012). Younger ages have been suggested for Nepal $\sim$15–11 Ma (Pearson and DeCelles 2005), $\sim$11–9 Ma, (Kohn et al. 2004) and for eastern Himachal Pradesh $\sim$10–0 Ma (Vannay et al. 2004), $\sim$10–6 Ma (Caddick et al. 2007). These differences may result from the different approaches to constraining the age of movement on a fault (e.g. thermochronological data, structural reconstructions) or from the inconsistent definition of the thrust(s) and the resulting difficulties in establishing along-strike correlations. In conclusion, although the MRST may be an orogen-scale structure traceable as far west as 75° E, it appears to be a complex structure consisting of several segments, the geometry and age of which vary along the orogen (Stübner et al. 2018).

### 3.3.4. *Main Frontal thrust (MFT)*

The MFT is a Quaternary fault located along the foot of the Himalaya, at the break in slope between the foothills and the alluvial plain of the foreland basin.

---

6. In fold-and-thrust belts, duplexes are bounded by a floor and roof thrust. Duplexes consist of a stack of horses; a horse is a panel of rock bounded on all sides by thrusts. A passive roof thrust is a thrust that takes no part in the displacement (or accommodating shortening). Such thrusts develop during underthrusting or wedge insertion. As a horse climbs over a ramp, it is inserted as a wedge whose tip is the leading branching point of the floor thrust and roof thrust. A passive roof thrust grows by the in-sequence addition of horses and abandonment of the previous horse tip.

It places the Siwaliks sediments on top of undeformed Quaternary sediments of the Indus–Ganges–Brahmaputra plain (see Volume 2 – Chapter 8). The MFT is often covered by soil and lush vegetation because of subtropical climate conditions. Exposures of the MFT are extremely rare, but those exposures that do exist have well-defined scarps cutting river terraces and alluvial fans (Nakata 1989). More commonly, the geometry of the system is inferred from the geomorphology and structural geology of its hanging wall (Yeats et al. 1992). The MFT may also be a blind thrust, the frontal part of the Siwalik range is often occupied by anticlinal ridges and in many places by synformal depressions of intermontane basins called duns. When exposure conditions allow the detailed mapping of its trace, it is evident that the MFT is a discontinuous structure which consists of en échelon multi-kilometric segments connected by transfer faults or accommodation zones (Figure 3.10). The MFT probably did not mature to develop a throughgoing, continuous trace because of the relatively small displacement along it. Even so, seismic data and topographic features demonstrate that the MFT has already propagated horizontally dozens of kilometers southwards, into the foreland basin, mostly as a blind thrust.

The trace of the MFT composite structure is discontinuous and in West Bengal and Central Bhutan – within large re-entrants – involves at least three traces: one at the topographic break, the topographic frontal thrust (TFT) and two outboard of the orogen (Nakata 1972), which apparently affect only the sediments of the foreland basin. The intermediate trace is a north-directed back thrust, interpreted as an element of a juvenile triangle zone (Stockmal et al. 2001) at the Himalayan orogenic front (Dasgupta et al. 2013; Chakrabarti Goswami et al. 2019). The southern tip of the MFT in the foreland basin may also be a blind basal décollement with an incipient south-dipping back thrust (Duvall et al. 2020). Although the first-order trace of the MFT (at the topographic break) appears curved but continuous, detailed mapping indicates that the trace consists of segments arranged in a left-stepping en échelon geometry (Figure 3.10). These segments are separated by strike-slip faults, pressure ridges or relay ramps, characteristic of fault growth by linkage, extensively investigated in extensional tectonic settings (e.g. Mansfield and Cartwright 2001).

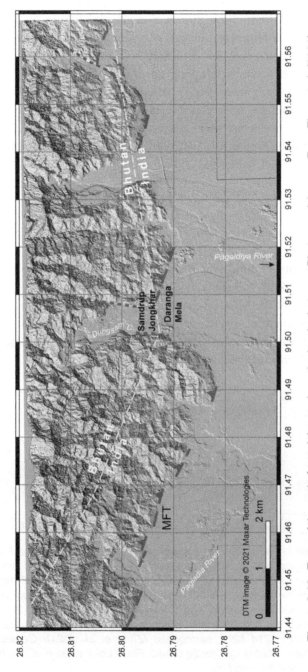

**Figure 3.10.** Trace of the main frontal thrust along the border between Bhutan and Assam. From Zhao et al. (2021)

Geomorphic analyses of river terraces and geodesy (GPS observations) (Lavé and Avouac 2000) have shown that the MFT was the only active structure in the sub-Himalaya during the Holocene. Over the Holocene, it has absorbed on average $21 \pm 1.5$ mm/yr of N-S shortening, that is, thrusting on the MFT seems to taking up most of the shortening across the whole Himalaya. During interseismic periods, the MHT is locked, and elastic deformation accumulates until being released by large ($M_W$ 8) earthquakes. These earthquakes break the MHT up to the near surface at the front of the Himalayan foothills (Figure 3.11) and result in incremental activation of the MFT (see Volume 3 – Chapter 5).

Structural analyses and dating of paleo-earthquakes indicate that most of the displacement along the MFT has been achieved by seismic slip, rather than by a continuous, smooth creep (Avouac 2015). Exposures of deformed sediments are generally obtained by the excavation of trenches across scarps produced during the most recent movements on the MFT. Since the pioneering work by Nakata et al. (1998), there were more than 30 studies along $\sim$ 2,000 km of the MFT (Wesnousky 2020) showing that the largest part of the MFT was ruptured by major earthquakes (Bilham 2019) since about 1,100 CE (Lavé et al. 2005), as well as that the entire Himalayan arc is poised to produce a sequence of great earthquakes (Wesnousky 2020).

**Figure 3.11.** *Main frontal thrust. (a) Natural outcrop along the river bank in eastern Bhutan. The hanging wall is made of Lower Siwalik mudstone and siltstone (ca. 7 Ma, Coutand et al. 2016). Tip of the hanging wall is to the left of the group of people. (b) Tip of the hanging wall. On the top Siwalik sandstone and siltstone. At the bottom sandy silt. Depositional age 1,458–1,638 CE (Zhao et al. 2021). Note a black layer between the two lithologies which is a fault gouge. This surface rupture was caused by the 1,714 M8.1 earthquake (Zhao et al. 2021). Geologist in the photo is Yuqiu Zhao. Photos by Djordje Grujic. For a color version of this figure, see www.iste.co.uk/cattin/himalaya1.zip*

### 3.3.5. *Main Himalayan thrust (MHT), continental megathrust*

The main frontal thrust system is the surface expression of a low-angle, basal thrust along which the Indian plate is underthrust beneath the Himalaya and southern Tibet and into which the MFT, MBT and MCT systems root (Molnar 1984; Ni and Barazangi 1984; Schelling and Arita 1991; Srivastava and Mitra 1994). In contrast, the STDS does not physically connect with those thrusts and hence is an entirely separate structure. This basal thrust was first imaged by natural seismicity data (Ni and Barazangi 1984), and observed by the INDEPTH seismic reflection project (Nelson et al. 1996), and named main Himalayan thrust (MHT). Since then, the structure was imaged in greater detail by natural seismicity data (Duputel et al. 2016; Elliott et al. 2016; Singer et al. 2017; Wang et al. 2017). INDEPTH geophysical results suggested that the MHT may extend North until the latitude of the Indus–Tsangpo suture zone, while receiver function data (Nábělek et al. 2009) indicate that this structure and the Indian crust may extend another 150–200 km further north where the Indian lower crust and lithospheric mantle become subducted (see Volume 1 – Chapter 4). This structure is therefore equivalent to an oceanic subduction zone, and represents a true, active continental megathrust, potentially capable of generating $M_W$ 9 earthquakes (Stevens and Avouac 2016).

The northern limit of seismic slip along the MHT is contained within the locked zone (Elliott et al. 2016), which is consistent with the generic, globally observed, behavior of active faults and megathrusts, in which seismic and aseismic portions appear mutually exclusive. As it is known from seismic studies in subduction zones, the stress along a megathrust is transferred from the deepest segments deforming by crystal plastic processes to the shallower segments deforming in a brittle-frictional manner and which are seismically active.

### 3.4. Tectonic models

### 3.4.1. *Fold-and-thrust belt versus channel flow*

The deformation style in the GHS and LHS is profoundly different. The GHS is pervasively ductily sheared and folded. The high-temperature conditions of deformation > 650°C of dominantly quartzo-felspathic and hydrous mineral bulk composition rocks lead to syntectonic anatexis and pervasive yet heterogeneous ductile shearing. Although the kilometer scale recumbent tight to isoclinal folds has been observed (Figure 3.12), there are no field observations suggesting that the GHS is a fold nappe, a recumbent, south facing antiform (e.g. Heim and Gansser 1939) in the sense of Alpine tectonics terminology. Folds are tight recumbent, with axial planes parallel to the pervasive foliation and fold hinges parallel to the stretching and mineral lineation.

**Figure 3.12.** *Jomolhari (also Chomolhari; 7326 m) on the western border of Bhutan. In the middle of its 2,800 m-high east face, there is a recumbent fold in the GHS gneisses. The snow-capped peak is built from the TSS rocks. Photo by Djordje Grujic. For a color version of this figure, see www.iste.co.uk/cattin/himalaya1.zip*

There are numerous structures indicating coeval anatexis and ductile deformation (Figure 3.13). In summary, under peak temperature deformation conditions, the GHS was not a rigid, deformable thrust "sheet". The peak metamorphic temperatures in the LHS rocks were at or below the brittle–ductile transition for quartzo-felspathic rocks. The deformation was much more localized into brittle–ductile shear zones and faults, deforming the outer LHS in the style of fold-and-thrust belts with up to about a dozen horses. The fold hinges are perpendicular to the transport direction and the axial planes moderately to steeply dipping. The sedimentary bedding is preserved and mostly the right way-up.

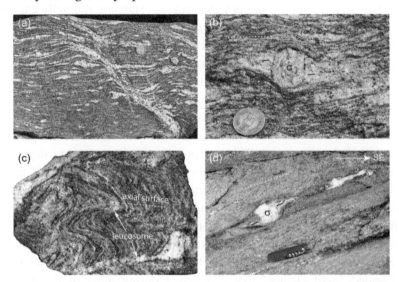

**Figure 3.13.** *Coeval partial melting and deformation in the GHS. (a) Migmatite with shear bands filled with leucosome. Photo by Roberto Weinberg. (b) σ-type feldspar porphyroclasts with asymmetric pressure shadows filled with leucosome. Photo by Roberto Weinberg. (c) Folded migmatite with leucosome veins parallel to the fold axial surfaces. Photo by Djordje Grujic. (d) Boudins of a leucogranite dyke with asymmetric pressure shadows filled with leucosome. Photo by Djordje Grujic. For a color version of this figure, see www.iste.co.uk/cattin/himalaya1.zip*

The MCT therefore separates two lithotectonic units with fundamentally different deformation styles: the upper unit is the GHS, characterized by dominantly horizontal flow, extending flow (i.e. extension in the direction of flow), whereas deformation in its lower portion was characterized by compressing flow (i.e. compression in the direction of flow) (Price 1972; Culshaw et al. 2006). Extending flow is a distinctive feature of displacement and distortion in deep orogenic hinterlands, while compressing flow is the emblematic of displacement and distortion in orogenic foreland regions. The transition between the upper and lower parts of the Himalaya therefore represents a transition between hinterland-style deformation, involving processes such as lateral midcrustal flow, and foreland-style deformation of fold-and-thrust belts. In the Himalaya, the former is represented by the channel flow (Grujic et al. 1996; Beaumont et al. 2004; Jamieson et al. 2004; Beaumont et al. 2006; Grujic 2006) and the latter by critical-taper (Coulomb) wedge development (Dahlen et al. 1984; Dahlen 1990; Hilley and Strecker 2004).

The two tectonic modes are not alternatives. They are mutually exclusive as they include fundamentally different deformation mechanisms. Critical Coulomb wedge deformation incorporates a brittle-frictional rheology and pore–fluid pressure effect (Dahlen 1990). Channel flow incorporates thermally activated viscous rheology and lateral pressure gradient (Beaumont et al. 2004). In a large hot orogen like the Himalaya, the internal part at mid-crustal levels is deforming according to the channel flow principles. In a tectonically thickened crust, the radiogenic heat production increases the temperatures eventually triggering anatexis. This causes a dramatic drop in viscosity and the onset of return flow of the channel. As the channel advances towards foreland, it cools and the flow component governed by the pressure gradient between the hinterland and foreland, between thick and normal crust, progressively vanishes. Further ductile deformation is governed by shear stresses. At the brittle–ductile transition, frictional deformation takes over. In the Himalaya, the GHS and probably the base of the TSS and the top of the LHS constitute the exhumed fossil channel that was active during most of the Miocene. The transition to viscous shearing occurred at temperatures below solidus. The current brittle–ductile transition along the MHT is located at the base of the seismogenic crust, at depths of about 10–15 km and temperatures below $\sim 350°C$ (Avouac 2015). The Himalayan orogen therefore currently deforms as a brittle–ductile wedge. It is not known if there is an active crustal channel and if so, where it is located. Some geophysical data suggest that the southern tip of it may be located underneath southern Tibet (Nelson et al. 1996; Klemperer 2006).

### 3.4.2. *Coeval slip along the STDS and the MCT*

Most of us tend to assume that thrust faults occur in contractional regimes and normal faults in extensional regimes. How could then the STDS with a slip exceeding 100 km form in a continental collisional zone without any evidence of extension in the N-S direction?

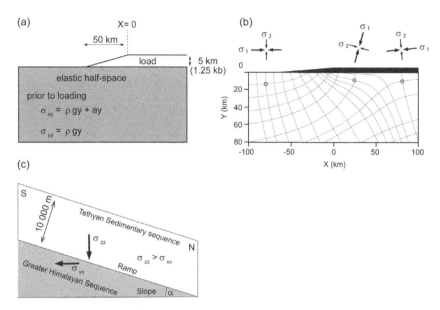

**Figure 3.14.** *Stress distribution in the Himalaya and Tibet. (a) Simple elastic model used to calculate the stress trajectories in (b). Indian lithosphere (gray) is taken to be an infinite elastic sheet. Load by Himalaya and Tibet increases linearly to a maximum 5 km, corresponding to 1.25 kbar. (b) Calculated stress trajectories. Note the rotation of principal stress axes near the southern edge of the plateau interpreted to cause normal faulting in the region. After Burchfiel and Royden (1985). (c) Northward normal faulting explained by a gravity-driven décollement of the Tethyan sedimentary pile from the crystalline basement (i.e. GHS). After Burg et al. (1984)*

The first geologists to study the STDS proposed that the gravitational load modifies the principal stress orientation to cause normal faulting along the crest of the Himalaya (Figure 3.14). This process could easily explain the moderately to steeply dipping normal faults, but not a low angle detachment. In the 1980s, this was a general conundrum in orogen understanding.

Channel flow (Turcotte and Schubert 2002; Beaumont et al. 2004; Grujic 2006) with the strong backflow phenomenon (relative to the lower plate; Figure 3.15) leads to synconvergent exhumation of the channel material and as a consequence thrusting motion at the base and normal shear at the top. From the basic governing equations, it is evident that the channel flow is sensitive to the channel material viscosity (function of internal heat production and rock type), pressure gradient (function of crustal thickness and therefore influenced by the surface denudation as well), channel thickness and convergence velocity. None of these parameters can be homogeneous in space or steady in time. Therefore, multiple pulses of channel flow along one transect or variable flow velocities along strike cause internal shear zones and lateral discontinuities.

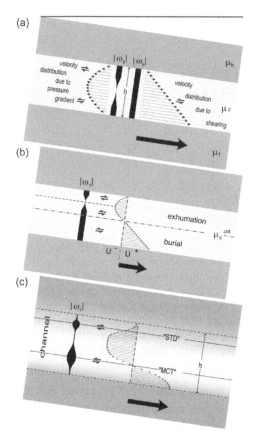

**Figure 3.15.** *Diagram of the flow pattern in a channel with $\mu_h > \mu_c > \mu_f$. The velocity reference frame is attached to the hanging wall. The vorticity values are indicated by the width of the black bar. Only the absolute value of the vorticity $|\omega|$ indicated regardless of whether it is positive (sinistral simple shear) or negative (dextral simple shear). (a) End-members of flow in a channel. Right, velocity profile caused by shearing, Couette flow. Left, velocity profile caused by pressure gradient within the channel. (b) There is a critical viscosity of the channel material below which the Poiseuille flow will counteract the shear forces and cause return flow (negative velocity, $U^-$) and therefore exhumation of that part of the channel material. That part of the channel that remains dominated by the induced shear (positive velocities, $U^+$) will continue being underthrust. (c) A hypothetical velocity profile in a Himalayan-type orogenic channel. Darker gray shades depict higher viscosities. Major structures like the MCT and the STD are located within the channel (Grujic 2006)*

## 3.5. Conclusion

It is important to understand that as an orogen evolves, the geodynamic conditions change and thus tectonic modes vary in time and space (along strike and with depth). Time and length scales of these changes are different for different geologic processes. Therefore, despite 50 Myr of evolution, Himalayan structures have not achieved a steady state.

On the other hand, despite more than 150 years of geological exploration, geologists did not yet meet a consensus even about the nature and consequently the location of the first-order Himalayan structures. Although efforts have been made to map the first-order shear zones and faults as continuous along the entire length of the orogen, there is growing evidence that these structures are segmented both at the surface and at depth. This segmentation and still patchy geochronological data may be one of the causes for inconsistencies in their definition. The resulting multitude of names, and inconsistent use of terminology, hampers fundamental advances in our understanding of the tectonic evolution of this orogen. Systematic work on deformation conditions, mechanisms and timing, and a consistent structural terminology are all needed to advance our quantitative understanding of how and when those main tectonic structures formed, developed and configured the Himalayan orogen.

## 3.6. References

Arita, K., Ohta, Y., Akiba, C., Maruo, Y. (1973). Kathmandu region. In *Geology of the Nepal Himalayas*, Hashimoto, S.E.A. (ed.). Saikon, Sapporo.

Avouac, J.P. (2015). Mountain building: From earthquakes to geologic deformation. In *Treatise on Geophysics*, Schubert, G. (ed.). Elsevier, Oxford.

Beaumont, C., Jamieson, R.A., Nguyen, M.H., Medvedev, S. (2004). Crustal channel flows 1: Numerical models with applications to the tectonics of the Himalayan-Tibetan orogen. *Journal of Geophysical Research-Solid Earth*, 109(B6), B06406.

Beaumont, C., Nguyen, M.H., Jamieson, R.A., Ellis, S. (2006). Crustal flow modes in large hot orogens. In *Channel Flow, Ductile Extrusion and Exhumation in Continental Collision Zones*, Law, R.D., Searle, M.P., Godin, L. (eds). The Geological Society, Special Publications, London.

van der Beek, P., Robert, X., Mugnier, J.-L., Bernet, M., Huyghe, P., Labrin, E. (2006). Late Miocene – Recent exhumation of the central Himalaya and recycling in the foreland basin assessed by apatite fission-track thermochronology of Siwalik sediments, Nepal. *Basin Research*, 18(4), 413–434. http://doi.wiley.com/10.1111/j.1365-2117.2006.00305.x.

Benetti, B., Montomoli, C., Iaccarino, S., Langone, A., Carosi, R. (2021). Mapping tectono-metamorphic discontinuities in orogenic belts: Implications for mid-crust exhumation in NW Himalaya. *Lithos*, 392, 106129.

Bilham, R. (2019). Himalayan earthquakes: A review of historical seismicity and early 21st century slip potential. *Geological Society, London, Special Publications*, 483(1), 423–482.

Bordet, P. (1961). *Recherches géologiques dans l'Himalaya du Népal, région du Makalu*. Centre national de la recherche scientifique, Paris.

Boyer, S.E. and Elliott, D. (1982). Thrust systems. *Bulletin of the American Association of Petroleum Geologists*, 66, 1196–1230.

Braun, J. (2016). Strong imprint of past orogenic events on the thermochronological record. *Tectonophysics*, 683, 325–332.

Brun, J.-P., Burg, J.-P., Ming, C. G. (1985). Strain trajectories above the Main Central Thrust (Himalaya) in southern Tibet. *Nature*, 313, 388–390.

Brunel, M. (1986). Ductile thrusting in the Himalayas: Shear sense criteria and stretching lineations. *Tectonics*, 5(2), 247–265.

Burchfiel, B.C. and Royden, L.H. (1985). North-south extension within the convergent Himalayan region. *Geology*, 13, 679–682.

Burchfiel, B.C., Chen, Z., Hodges, K.V., Liu, Y., Royden, L.H., Deng, C., Xu, J. (1992). The South Tibetan detachment system, Himalayan orogen: Extension contemporaneous with and parallel to shortening in a collisional mountain belt. *Geological Society of America Special Paper*, 269, 41.

Burg, J.P., Brunel, M., Gapais, D., Chen, G.M., Liu, G.H. (1984). Deformation of leucogranites of the crystalline Main Central Sheet in southern Tibet. *Journal of Structural Geology*, 6, 535–542.

Caddick, M.J., Bickle, M.J., Harris, N.B.W., Holland, T.J.B., Horstwood, M.S.A., Parrish, R.R., Ahmad, T. (2007). Burial and exhumation history of a Lesser Himalayan schist: Recording the formation of an inverted metamorphic sequence in NW India. *Earth and Planetary Science Letters*, 264(3–4), 375–390.

Campani, M., Herman, F., Mancktelow, N. (2010). Two-and three-dimensional thermal modeling of a low-angle detachment: Exhumation history of the Simplon Fault Zone, central Alps. *Journal of Geophysical Research: Solid Earth*, 115, B10420.

Carosi, R., Lombardo, B., Molli, G., Musumeci, G., Pertusati, P.C. (1998). The south Tibetan detachment system in the Rongbuk valley, Everest region. Deformation features and geological implications. *Journal of Asian Earth Sciences*, 16(2–3), 299–311.

Carosi, R., Montomoli, C., Rubatto, D., Visonà, D. (2006). Normal-sense shear zones in the core of the Higher Himalayan Crystallines (Bhutan Himalaya): Evidence for extrusion? *Geological Society, London, Special Publications*, 268(1), 425–444.

Carosi, R., Montomoli, C., Iaccarino, S. (2018). 20 years of geological mapping of the metamorphic core across Central and Eastern Himalayas. *Earth-Science Reviews*, 177, 124–138.

Cawood, P.A., Johnson, M.R.W., Nemchin, A.A. (2007). Early Palaeozoic orogenesis along the Indian margin of Gondwana: Tectonic response to Gondwana assembly. *Earth and Planetary Science Letters*, 255(1–2), 70–84.

Chakrabarti Goswami, C., Jana, P., Weber, J.C. (2019). Evolution of landscape in a piedmont section of Eastern Himalayan foothills along India–Bhutan border: A tectono-geomorphic perspective. *Journal of Mountain Science*, 16(12), 2828–2843.

Chakraborty, S., Anczkiewicz, R., Gaidies, F., Rubatto, D., Sorcar, N., Faak, K., Mukhopadhyay, D., Dasgupta, S. (2016). A review of thermal history and timescales of tectonometamorphic processes in Sikkim Himalaya (NE India) and implications for rates of metamorphic processes. *Journal of Metamorphic Geology*, 34(8), 785–803.

Chakungal, J., Dostal, J., Grujic, D., Duchêne, S., Ghalley, S.K. (2010). Provenance of the Greater Himalayan Sequence: Evidence from mafic eclogite-granulites and amphibolites in NW Bhutan. *Tectonophysics*, 480, 198–212.

Colchen, M., Le, F.P., Pêcher, A. (1986). *Recherches géologiques dans l'Himalaya du Népal : Annapurna-Manaslu-Ganesh Himal*. Editions du Centre National de la Recherche Scientifique, Paris.

Cottle, J.M., Jessup, M.J., Newell, D.L., Horstwood, M.S.A., Noble, S.R., Parrish, R.R., Waters, D.J., Searle, M.P. (2009). Geochronology of granulitized eclogite from the Ama Drime Massif: Implications for the tectonic evolution of the South Tibetan Himalaya. *Tectonics*, 28(TC1002). doi:10.1029/2008TC002256.

Cottle, J., Waters, D., Riley, D., Beyssac, O., Jessup, M. (2011). Metamorphic history of the South Tibetan Detachment System, Mt. Everest region, revealed by RSCM thermometry and phase equilibria modelling. *Journal of Metamorphic Geology*, 29(5), 561–582.

Coutand, I., Barrier, L., Govin, G., Grujic, D., Dupont-Nivet, G., Najman, Y., Hoorn, C. (2016). Late Miocene–Pleistocene evolution of India–Eurasia convergence partitioning between the Bhutan Himalaya and the Shillong plateau: New evidences from foreland basin deposits along the Dungsam Chu section, Eastern Bhutan. *Tectonics*, 35(12), 2963–2994.

Culshaw, N., Beaumont, C., Jamieson, R. (2006). The orogenic superstructure-infrastructure concept: Revisited, quantified, and revived. *Geology*, 34(9), 733–736.

Dahlen, F.A. (1990). Critical taper model of fold-and-thrust belts and accretionary wedges. *Annual Review of Earth and Planetary Sciences*, 18, 55–99.

Dahlen, F.A., Suppe, J., Davis, D. (1984). Mechanics of fold-and-thrust belts and accretionary wedges: Cohesive Coulomb Theory. *Journal of Geophysical Research*, 89(B12), 10087.

Dahlstrom, C.D. (1970). Structural geology in the eastern margin of the Canadian Rocky Mountains. *Bulletin of Canadian Petroleum Geology*, 18(3), 332–406.

Dasgupta, S., Mazumdar, K., Moirangcha, L., Gupta, T.D., Mukhopadhyay, B. (2013). Seismic landscape from Sarpang re-entrant, Bhutan Himalaya foredeep, Assam, India: Constraints from geomorphology and geology. *Tectonophysics*, 592, 130–140.

Duputel, Z., Vergne, J., Rivera, L., Wittlinger, G., Farra, V., Hetényi, G. (2016). The 2015 Gorkha earthquake: A large event illuminating the Main Himalayan Thrust fault. *Geophysical Research Letters*, 43(6), 2517–2525.

Duvall, M.J., Waldron, J.W., Godin, L., Najman, Y. (2020). Active strike-slip faults and an outer frontal thrust in the Himalayan foreland basin. *Proceedings of the National Academy of Sciences*, 117(30), 17615–17621.

Elliott, J., Jolivet, R., González, P.J., Avouac, J.-P., Hollingsworth, J., Searle, M., Stevens, V. (2016). Himalayan megathrust geometry and relation to topography revealed by the Gorkha earthquake. *Nature Geoscience*, 9(2), 174–180.

England, P. and Molnar, P. (1990). Surface uplift, uplift of rocks, and exhumation of rocks. *Geology*, 18(12), 1173.

Epard, J.-L. and Steck, A. (2004). The Eastern prolongation of the Zanskar Shear Zone (Western Himalaya). *Eclogae Geologicae Helvetiae (Swiss Journal of Geosciences)*, 97(2), 193–212.

Ferrara, G., Lombardo, B., Tonarini, S. (1983). Rb/Sr geochronology of granites and gneisses from the Mount Everest region, Nepal Himalaya. *Geologische Rundschau*, 72(1), 119–136.

Finch, M., Hasalová, P., Weinberg, R.F., Fanning, C.M. (2014). Switch from thrusting to normal shearing in the Zanskar shear zone, NW Himalaya: Implications for channel flow. *GSA Bulletin*, 126(7–8), 892–924.

Fossen, H. and Cavalcante, G.C.G. (2017). Shear zones – A review. *Earth-Science Reviews*, 171, 434–455.

Ganguly, J., Dasgupta, S., Cheng, W., Neogi, S. (2000). Exhumation history of a section of the Sikkim Himalayas, India: Records in the metamorphic mineral equilibria and compositional zoning of garnet. *Earth and Planetary Science Letters*, 183(3–4), 471–486.

Gansser, A. (1964). *Geology of the Himalayas*. Interscience div. of John Wiley & Sons Ltd, New York, Sydney.

Gansser, A. (1983). *Geology of the Bhutan Himalaya*. Birkhäuser Verlag, Basel, Boston, Stuttgart.

Godin, L., Ahenda, M., Grujic, D., Stevenson, R., Cottle, J. (2021). Protolith affiliation and tectonometamorphic evolution of the Gurla Mandhata core complex, NW Nepal Himalaya. *Geosphere*, 17(2), 626–646.

Goscombe, B., Gray, D., Hand, M. (2006). Crustal architecture of the Himalayan metamorphic front in eastern Nepal. *Gondwana Research*, 10(3–4), 232–255.

Goscombe, B., Gray, D., Foster, D.A. (2018). Metamorphic response to collision in the Central Himalayan Orogen. *Gondwana Research*, 57, 191–265.

Grujic, D. (2006). Channel flow and continental collision tectonics: An overview. In *Channel Flow, Ductile Extrusion and Exhumation in Continental Collision Zones*, Law, R.D., Searle, M.P., Godin, L. (eds). Geological Society, Special Publications, London.

Grujic, D., Casey, M., Davidson, C., Hollister, L.S., Kündig, R., Pavlis, T., Schmid, S. (1996). Ductile extrusion of the Higher Himalayan Crystalline in Bhutan: Evidence from quartz microfabrics. *Tectonophysics*, 260(1–3), 21–43.

Grujic, D., Hollister, L., Parrish, R. (2002). Himalayan metamorphic sequence as an orogenic channel: Insight from Bhutan. *Earth and Planetary Science Letters*, 198(1–2), 177–191.

Grujic, D., Warren, C.J., Wooden, J.L. (2011). Rapid synconvergent exhumation of Miocene-aged lower orogenic crust in the eastern Himalaya. *Lithosphere*, 3(5), 346–366.

Grujic, D., Ashley, K.T., Coble, M.A., Coutand, I., Kellett, D.A., Larson, K.P., Whipp Jr., D.M., Gao, M., Whynot, N. (2020). Deformational temperatures across the lesser himalayan sequence in eastern Bhutan and their implications for the deformation history of the Main Central Thrust. *Tectonics*, 39(4), e2019TC005914.

Hamet, J. and Allègre, C.-J. (1976). Rb-Sr systematics in granite from central Nepal (Manaslu): Significance of the Oligocene age and high 87Sr/86Sr ratio in Himalayan orogeny. *Geology*, 4(8), 470–472.

Heim, A. and Gansser, A. (1939). Central Himalaya: Geological observations of the Swiss expedition, 1936. *Denkschriften der Schweizerischen Naturforschenden Gesellschaft*, 73(1), 245.

Herren, E. (1987). Zanskar shear zone: Northeast-southwest extension within the higher Himalayas (Ladakh, India). *Geology*, 15(5), 409–413.

Hilley, G. and Strecker, M. (2004). Steady state erosion of critical Coulomb wedges with applications to Taiwan and the Himalaya. *Journal of Geophysical Research*, 109, B01411.

Hunter, N.J., Weinberg, R.F., Wilson, C.J., Luzin, V., Misra, S. (2018). Microscopic anatomy of a "hot-on-cold" shear zone: Insights from quartzites of the Main Central Thrust in the Alaknanda region (Garhwal Himalaya). *Bulletin*, 130(9–10), 1519–1539.

Hurtado, J.M.J., Hodges, K.V., Whipple, K.X. (2001). Neotectonics of the Thakkhola graben and implications for recent activity on the South Tibetan fault system in the central Nepal Himalaya. *Geological Society of America Bulletin*, 113(2), 222–240.

Jamieson, R.A., Beaumont, C., Medvedev, S., Nguyen, M.H. (2004). Crustal channel flows 2: Numerical models with implications for metamorphism in the Himalayan-Tibetan orogen. *Journal of Geophysical Research*, 109(B06407). doi:10.1029/2003JB002811.

Jamieson, R.A., Beaumont, C., Nguyen, M., Grujic, D. (2006). Provenance of the Greater Himalayan Sequence and associated rocks: Predictions of channel flow models. *Geological Society London Special Publications*, 268, 165.

Kellett, D.A. and Grujic, D. (2012). New insight into the South Tibetan detachment system: Not a single progressive deformation. *Tectonics*, 31(2), TC2007.

Kellett, D.A., Grujic, D., Erdmann, S. (2009). Miocene structural reorganization of the South Tibetan detachment, eastern Himalaya: Implications for continental collision. *Lithosphere*, 1(5), 259–281.

Kellett, D.A., Grujic, D., Coutand, I., Cottle, J., Mukul, M. (2013). The South Tibetan detachment system facilitates ultra rapid cooling of granulite-facies rocks in Sikkim Himalaya. *Tectonics*, 32(2), 252–270.

Kellett, D.A., Cottle, J.M., Larson, K.P. (2019). The South Tibetan Detachment System: History, advances, definition and future directions. *Geological Society, London, Special Publications*, 483(1), 377–400.

Klemperer, S.L. (2006). Crustal flow in Tibet: Geophysical evidence for the physical state of Tibetan lithosphere, and inferred patterns of active flow. In *Channel Flow, Ductile Extrusion and Exhumation in Continental Collision Zones*, Godin, L., Grujic, D., Law, R.D., Searle, M.P. (eds). The Geological Society, London.

Kohn, M.J., Wieland, M.S., Parkinson, C.D., Upreti, B.N. (2004). Miocene faulting at plate tectonic velocity in the Himalaya of central Nepal. *Earth and Planetary Science Letters*, 228(3–4), 299–310.

Kohn, M.J., Paul, S.K., Corrie, S.L. (2010). The lower Lesser Himalayan sequence: A Paleoproterozoic arc on the northern margin of the Indian plate. *Bulletin*, 122(3–4), 323–335.

Kündig, R. (1988). Kristallisation und deformation im higher Himalaya, Zanskar (NW-Indien). Unpublished PhD Thesis, ETH Zürich.

La Roche, R.S., Godin, L., Crowley, J.L. (2018). Reappraisal of emplacement models for Himalayan external crystalline nappes: The Jajarkot klippe, western Nepal. *Geological Society of America Bulletin*, 130(5–6), 1041–1056.

Larson, K.P. and Godin, L. (2009). Kinematics of the Greater Himalayan sequence, Dhaulagiri Himal: Implications for the structural framework of central Nepal. *Journal of the Geological Society*, 166(1), 25–43.

Larson, K.P., Cottle, J., Lederer, G., Rai, S. (2017). Defining shear zone boundaries using fabric intensity gradients: An example from the east-central Nepal Himalaya. *Geosphere*, 13(3), 771–781.

Lavé, J. and Avouac, J. (2000). Active folding of fluvial terraces across the Siwaliks Hills, Himalayas of central Nepal. *Journal of Geophysical Research-Solid Earth*, 105(B3), 5735–5770.

Lavé, J., Yule, D., Sapkota, S., Basant, K., Madden, C., Attal, M., Pandey, R. (2005). Evidence for a great medieval earthquake ($\tilde{1}100$ AD) in the central Himalayas, Nepal. *Science*, 307(5713), 1302–1305.

Law, R., Jessup, M., Searle, M., Francsis, M., Waters, D., Cottle, J. (2011). Telescoping of isotherms beneath the South Tibetan detachment system, Mount Everest Massif. *Journal of Structural Geology*, 33(11), 1569–1594.

Law, R., Stahr III, D., Francsis, M., Ashley, K., Grasemann, B., Ahmad, T. (2013). Deformation temperatures and flow vorticities near the base of the Greater Himalayan Series, Sutlej Valley and Shimla Klippe, NW India. *Journal of Structural Geology*, 54, 21–53.

Le Fort, P. (1975). Himalayas: The collided range. Present knowledge of the continental arc. *American Journal of Science*, 275-A, 1–44.

Leloup, P.H., Mahéo, G., Arnaud, N., Kali, E., Boutonnet, E., Liu, D., Xiaohan, L., Haibing, L. (2010). The South Tibet detachment shear zone in the Dinggye area: Time constraints on extrusion models of the Himalayas. *Earth and Planetary Science Letters*, 292, 1–16.

Long, S.P., McQuarrie, N., Tobgay, T., Grujic, D. (2011a). Geometry and crustal shortening of the Himalayan fold-thrust belt, eastern and central Bhutan. *Geological Society of America Bulletin*, 123(7–8), 1427–1447.

Long, S.P., McQuarrie, N., Tobgay, T., Hawthorne, J. (2011b). Quantifying internal strain and deformation temperature in the eastern Himalaya, Bhutan: Implications for the evolution of strain in thrust sheets. *Journal of Structural Geology*, 33(4), 579–608.

Long, S.P., McQuarrie, N., Tobgay, T., Rose, C., Gehrels, G., Grujic, D. (2011c). Tectonostratigraphy of the Lesser Himalaya of Bhutan: Implications for the along-strike stratigraphic continuity of the northern Indian margin. *Geological Society of America Bulletin*, 123(7–8), 1406–1426.

Long, S.P., McQuarrie, N., Tobgay, T., Coutand, I., Cooper, F.J., Reiners, P.W., Wartho, J.-A., Hodges, K.V. (2012). Variable shortening rates in the eastern Himalayan thrust belt, Bhutan: Insights from multiple thermochronologic and geochronologic data sets tied to kinematic reconstructions. *Tectonics*, 31(5).

Long, S.P., Gordon, S.M., Young, J.P., Soignard, E. (2016). Temperature and strain gradients through Lesser Himalayan rocks and across the Main Central thrust, south central Bhutan: Implications for transport-parallel stretching and inverted metamorphism. *Tectonics*, 35(8), 1863–1891.

Long, S.P., Mullady, C.L., Starnes, J.K., Gordon, S.M., Larson, K.P., Pianowski, L.S., Miller, R.B., Soignard, E. (2019). A structural model for the South Tibetan detachment system in northwestern Bhutan from integration of temperature, fabric, strain, and kinematic data. *Lithosphere*, 11(4), 465–487.

Mallet, F.R. (1875). On the geology and mineral resources of the Darjiling District in the Western Duars. *Memoirs of the Geological Survey of India*, XI, 1–50.

Mancktelow, N.S. (1990). The Simplon fault zone. *Beiträge zur geologischen Karte der Schweiz*, n.F., 163.

Mandal, S., Robinson, D.M., Kohn, M.J., Khanal, S., Das, O. (2019). Examining the tectono-stratigraphic architecture, structural geometry, and kinematic evolution of the Himalayan fold-thrust belt, Kumaun, northwest India. *Lithosphere*, 11(4), 414–435.

Mansfield, C. and Cartwright, J. (2001). Fault growth by linkage: Observations and implications from analogue models. *Journal of Structural Geology*, 23(5), 745–763.

McQuarrie, N., Eizenhöfer, P.R., Long, S.P., Tobgay, T., Ehlers, T.A., Blythe, A.E., Morgan, L.E., Gilmore, M.E., Dering, G.M. (2019). The influence of foreland structures on hinterland cooling: Evaluating the drivers of exhumation in the eastern Bhutan Himalaya. *Tectonics*, 38(9), 3282–3310.

Meigs, A.J., Burbank, D.W., Beck, R.A. (1995). Middle-late Miocene (>10 Ma) formation of the Main Boundary thrust in the western Himalaya. *Geology*, 23(5), 423–426 [Online]. Available at: http://geology.gsapubs.org/content/23/5/423.short.

Meyer, M.C., Wiesmayr, G., Brauner, M., Häusler, H., Wangda, D. (2006). Active tectonics in Eastern Lunana (NW Bhutan): Implications for the seismic and glacial hazard potential of the Bhutan Himalaya. *Tectonics*, 25, TC3001. doi:3010.1029/2005TC001858.

Molnar, P. (1984). Structure and tectonics of the Himalaya; constraints and implications of geophysical data. *Annual Review of Earth and Planetary Sciences*, 12, 489–518.

Montomoli, C., Iaccarino, S., Carosi, R., Langone, A., Visonà, D. (2013). Tectonometamorphic discontinuities within the Greater Himalayan Sequence in Western Nepal (Central Himalaya): Insights on the exhumation of crystalline rocks. *Tectonophysics*, 608, 1349–1370.

Montomoli, C., Carosi, R., Rubatto, D., Visonà, D., Iaccarino, S. (2017). Tectonic activity along the inner margin of the South Tibetan Detachment constrained by syntectonic Leucogranite emplacement in Western Bhutan. *Italian Journal of Geosciences*, 136(1), 5–14.

Mugnier, J.-L., Huyghe, P., Chalaron, E., Mascle, G. (1994). Recent movements along the Main Boundary Thrust of the Himalayas: Normal faulting in an over-critical thrust wedge? *Tectonophysics*, 238(1–4), 199–215.

Mugnier, J.-L., Huyghe, P., Leturmy, P., Jouanne, F. (2004). Episodicity and rates of thrust sheet motion in Himalaya (Western Nepal). In *Thrust Tectonics and Hydrocarbon Systems*, McClay, K.R. (ed.). American Association of Petroleum Geologists Memoir, Tulsa.

Nábělek, J., Hetényi, G., Vergne, J., Sapkota, S., Kafle, B., Jiang, M., Su, H., Chen, J., Huang, B.S., Team, T.H.C. (2009). Underplating in the Himalaya-Tibet collision zone revealed by the Hi-CLIMB experiment. *Science*, 325(5946), 1371–1374.

Nakata, T. (1972). Geomorphic history and crustal movement of the foot-hills of the Himalayas. *The Science Reports of the Tohoku University 7th Series (Geography)*, 22(1), 39–177.

Nakata, T. (1989). Active faults of the Himalaya of India and Nepal. *Geological Society of America Special Paper*, 232(1), 243–264.

Nakata, T., Kumura, K., Rockwell, T. (1998). First successful paleoseismic trench study on active faults in the Himalaya. *Eos Transactions American Geophysical Union*, 79(45), 459–486.

Nelson, K.D., Zhao, W., Brown, L.D., Kuo, J., Che, J., Liu, X., Klemperer, S.L., Makovsky, Y., Meissner, R.J.J.M., Mechie, J., Kind, R. (1996). Partially molten middle crust beneath southern Tibet; synthesis of Project INDEPTH results. *Science*, 274(5293), 1684–1688.

Ni, J. and Barazangi, M. (1984). Seismotectonics of the Himalayan collision zone: Geometry of the underthrusting Indian plate beneath the Himalaya. *Journal of Geophysical Research: Solid Earth*, 89(B2), 1147–1163.

Oldham, R.D. (1893). *A Manual on the Geology of India, Stratigraphical and Structural Geology*, 2nd edition. Government of India, Kolkata.

Parsons, A., Law, R., Lloyd, G., Phillips, R., Searle, M. (2016). Thermo-kinematic evolution of the Annapurna-Dhaulagiri Himalaya, central Nepal: The composite orogenic system. *Geochemistry, Geophysics, Geosystems*, 17(4), 1511–1539.

Passchier, C.W. and Trouw, R.A. (2006). *Microtectonics*, 2nd edition. Springer-Verlag, Berlin, Heidelberg, New York.

Pearson, O.N. and DeCelles, P.G. (2005). Structural geology and regional tectonic significance of the Ramgarh thrust, Himalayan fold-thrust belt of Nepal. *Tectonics*, 24(4).

Pêcher, A. (1977). Geology of the Nepal Himalaya: Deformation and petrography in the Main Central Thrust zone. *Himalaya. Colloq. Int.*, 268.

Pêcher, A. (1978). Déformations et métamorphisme associés à une zone de cisaillement. Exemple du grand Chevauchement Central Himalayen (M.C.T.), transversale des Anapurnas et du Manaslu (Népal). PhD Thesis, Université Scientifique et Médicale de Grenoble.

Pêcher, A. (1991). The contact between the Higher Himalaya Crystallines and the Tibetan sedimentary series: Miocene large-scale dextral shearing. *Tectonics*, 10(3), 587–598.

Price, R.A. (1972). The distinction between displacement and distortion in flow, and the origin of diachronism in tectonic overprint in orogenic belts. *24th IGC*, Montreal.

Ramsay, J.G. (1980). Shear zone geometry: A review. *Journal of Structural Geology*, 2(1/2), 83–99.

Rana, N., Bhattacharya, F., Basavaiah, N., Pant, R., Juyal, N. (2013). Soft sediment deformation structures and their implications for Late Quaternary seismicity on the south Tibetan detachment system, central Himalaya (Uttarakhand), India. *Tectonophysics*, 592, 165–174.

Ratschbacher, L., Frisch, W., Liu, G., Chen, C. (1994). Distributed deformation in southern and western Tibet during and after the India-Asia collision. *Journal of Geophysical Research*, 99(B10), 19,917–919,945.

Riesner, M., Bollinger, L., Hubbard, J., Guérin, C., Lefèvre, M., Vallage, A., Basnet Shah, C., Kandel, T.P., Haines, S., Sapkota, S.N. (2021). Localized extension in megathrust hanging wall following great earthquakes in western Nepal. *Scientific Reports*, 11(1), 1–18.

Roberts, A.G., Weinberg, R.F., Hunter, N.J.R., Ganade, C.E. (2020). Large-scale rotational motion within the Main Central Thrust Zone in the Darjeeling-Sikkim Himalaya, India. *Tectonics*, 39, e2019TC005949.

Robinson, D.M. and Martin, A.J. (2014). Reconstructing the Greater Indian margin: A balanced cross section in central Nepal focusing on the Lesser Himalayan duplex. *Tectonics*, 33(11), 2143–2168.

Robinson, D.M. and McQuarrie, N. (2012). Pulsed deformation and variable slip rates within the central Himalayan thrust belt. *Lithosphere*, 4(5), 449–464.

Robinson, D.M. and Pearson, O.N. (2013). Was Himalayan normal faulting triggered by initiation of the Ramgarh–Munsiari thrust and development of the Lesser Himalayan duplex? *International Journal of Earth Sciences*, 102(7), 1773–1790.

Rubatto, D., Chakraborty, S., Dasgupta, S. (2012). Timescales of crustal melting in the Higher Himalayan Crystallines (Sikkim, Eastern Himalaya) inferred from trace element-constrained monazite and zircon chronology. *Contributions to Mineralogy and Petrology*, 165(2), 349–372.

Schelling, D. and Arita, K. (1991). Thrust tectonics, crustal shortening, and the structure of the far-eastern Nepal Himalaya. *Tectonics*, 10(5), 851–862.

Schmid, S., Zingg, A., Handy, M. (1987). The kinematics of movements along the Insubric Line and the emplacement of the Ivrea Zone. *Tectonophysics*, 135(1–3), 47–66.

Searle, M. (1999). Emplacement of Himalayan leucogranites by magma injection along giant sill complexes: Examples from the Cho Oyu, Gyachung Kang and Everest leucogranites (Nepal Himalaya). *Journal of Asian Earth Sciences*, 17(5–6), 773–783.

Searle, M., Simpson, R., Law, R., Parrish, R., Waters, D. (2003). The structural geometry, metamorphic and magmatic evolution of the Everest massif, High Himalaya of Nepal–South Tibet. *Journal of the Geological Society*, 160(3), 345–366.

Searle, M., Law, R.D., Godin, L., Larson, K.P., Streule, M.J., Cottle, J.M., Jessup, M.J. (2008). Defining the Himalayan Main Central Thrust in Nepal. *Journal of the Geological Society*, 165(2), 523–534.

Shurtleff, B.L. (2015). Rapid middle to late Miocene slip along the Zanskar normal fault, Greater Himalayan Range, NW, India: Constraints from low-temperature thermochronometry. MSc Master's Thesis, Central Washington University.

Singer, J., Obermann, A., Kissling, E., Fang, H., Hetényi, G., Grujic, D. (2017). Along-strike variations in the Himalayan orogenic wedge structure in Bhutan from ambient seismic noise tomography. *Geochemistry, Geophysics, Geosystems*, 18(4), 1483–1498.

Singh, P. and Patel, R. (2022). Miocene development of the Main Boundary Thrust and Ramgarh Thrust, and exhumation of Lesser Himalayan rocks of the Kumaun-Garhwal region, NW-Himalaya (India): Insights from fission track thermochronology. *Journal of Asian Earth Sciences*, 224, 104987.

Soucy La Roche, R., Godin, L., Cottle, J.M., Kellett, D.A. (2018). Preservation of the early evolution of the Himalayan middle crust in foreland klippen: Insights from the Karnali klippe, west Nepal. *Tectonics*, 37(5), 1161–1193.

Srivastava, P. and Mitra, G. (1994). Thrust geometries and deep structure of the outer and lesser Himalaya, Kumaon and Garhwal (India): Implications for evolution of the Himalayan fold-and-thrust belt. *Tectonics*, 13(1), 89–109.

Steck, A. (2003). Geology of the NW Indian Himalaya. *Eclogae Geologicae Helvetiae*, 96, 147–196.

Stevens, V. and Avouac, J.P. (2016). Millenary Mw> 9.0 earthquakes required by geodetic strain in the Himalaya. *Geophysical Research Letters*, 43(3), 1118–1123.

Stöcklin, J. (1980). Geology of Nepal and its regional frame. *Journal of the Geological Society*, 137(1), 1–34.

Stockmal, G.S., Lebel, D., McMechan, M.E., Mackay, P.A. (2001). Structural style and evolution of the triangle zone and external foothills, southwestern Alberta: Implications for thin-skinned thrust-and-fold belt mechanics. *Bulletin of Canadian Petroleum Geology*, 49(4), 472–496.

Stübner, K., Grujic, D., Parrish, R.R., Roberts, N.M., Kronz, A., Wooden, J., Ahmad, T. (2014). Monazite geochronology unravels the timing of crustal thickening in NW Himalaya. *Lithos*, 210, 111–128.

Stübner, K., Grujic, D., Dunkl, I., Thiede, R., Eugster, P. (2018). Pliocene episodic exhumation and the significance of the Munsiari thrust in the northwestern Himalaya. *Earth and Planetary Science Letters*, 481, 273–283.

Swapp, S.M. and Hollister, L.S. (1989). Thermobarometric constraints on deformation in the eastern High Himalaya, Bhutan. *Geological Society of America, 1989 Annual Meeting*, St. Louis, 6–9 November.

Thakur, V., Jayangondaperumal, R., Malik, M. (2010). Redefining Medlicott–Wadia's main boundary fault from Jhelum to Yamuna: An active fault strand of the main boundary thrust in northwest Himalaya. *Tectonophysics*, 489(1–4), 29–42.

Thiede, R., Robert, X., Stübner, K., Dey, S., Faruhn, J. (2017). Sustained out-of-sequence shortening along a tectonically active segment of the Main Boundary thrust: The Dhauladhar Range in the northwestern Himalaya. *Lithosphere*, 9(5), 715–725.

Turcotte, D.L. and Schubert, G. (2002). *Geodynamics*, 2nd edition. Cambridge.

Valdiya, K. (1980). The two intracrustal boundary thrusts of the Himalaya. *Tectonophysics*, 66(4), 323–348.

Vannay, J.C. and Hodges, K. (1996). Tectonometamorphic evolution of the Himalayan metamorphic core between the Annapurna and Dhaulagiri, central Nepal. *Journal of Metamorphic Geology*, 14(5), 635–656.

Vannay, J.C., Grasemann, B., Rahn, M., Frank, W., Carter, A., Baudraz, V., Cosca, M. (2004). Miocene to Holocene exhumation of metamorphic crustal wedges in the NW Himalaya: Evidence for tectonic extrusion coupled to fluvial erosion. *Tectonics*, 23(1).

Wang, X., Wei, S., Wu, W. (2017). Double-ramp on the Main Himalayan Thrust revealed by broadband waveform modeling of the 2015 Gorkha earthquake sequence. *Earth and Planetary Science Letters*, 473, 83–93.

Warren, C., Grujic, D., Kellett, D., Cottle, J., Jamieson, R.A., Ghalley, K. (2011). Probing the depths of the India-Asia collision: U-Th-Pb monazite chronology of granulites from NW Bhutan. *Tectonics*, 30(2).

Waters, D.J., Law, R.D., Searle, M.P., Jessup, M.J. (2019). Structural and thermal evolution of the South Tibetan Detachment shear zone in the Mt Everest region, from the 1933 sample collection of LR Wager. *Geological Society, London, Special Publications*, 478(1), 335–372.

Webb, A.A.G. (2013). Preliminary balanced palinspastic reconstruction of Cenozoic deformation across the Himachal Himalaya (northwestern India). *Geosphere*, 9(3), 572–587.

Webb, A.A.G., Yin, A., Harrison, T.M., Célérier, J., Gehrels, G.E., Manning, C.E., Grove, M. (2011). Cenozoic tectonic history of the Himachal Himalaya (northwestern India) and its constraints on the formation mechanism of the Himalayan orogen. *Geosphere*, 7(4), 1013–1061.

Wesnousky, S.G. (2020). Great pending Himalaya earthquakes. *Seismological Research Letters*, 91(6), 3334–3342.

Wiesmayr, G. and Grasemann, B. (2002). Eohimalayan fold and thrust belt: Implications for the geodynamic evolution of the NW-Himalaya (India). *Tectonics*, 21(6), 8-1–8-18.

Williams, C.A., Connors, C., Dahlen, F., Price, E.J., Suppe, J. (1994). Effect of the brittle-ductile transition on the topography of compressive mountain belts on Earth and Venus. *Journal of Geophysical Research: Solid Earth*, 99(B10), 19947–19974.

Yeats, R.S. and Lillie, R.J. (1991). Contemporary tectonics of the Himalayan frontal fault system: Folds, blind thrusts and the 1905 Kangra earthquake. *Journal of Structural Geology*, 13(2), 215–225.

Yeats, R.S., Nakata, T., Farah, A., Fort, M., Mirza, M.A., Pandey, M.R., Stein, R.S. (1992). The Himalayan Frontal Fault System. *Annales Tectonicae*, 6, 85–98.

Zhao, Y., Grujic, D., Baruah, S., Drukpa, D., Elkadi, J., Hetényi, G., King, G.E., Mildon, Z.K., Nepal, N., Welte, C. (2021). Paleoseismological findings at a new trench indicate the 1714 M8. 1 earthquake ruptured the main frontal thrust over all the Bhutan Himalaya. *Frontiers in Earth Science*, 9(9), 689457.

# PART 2

# Along Strike Variations

# 4

# Seismological Imaging and Current Seismicity of the Himalayan Arc

György HETÉNYI[1], Jérôme VERGNE[2], Laurent BOLLINGER[3],
Shiba SUBEDI[1,4], Konstantinos MICHAILOS[1]
and Dowchu DRUKPA[5]

[1] *Institute of Earth Sciences, University of Lausanne, Switzerland*
[2] *University of Strasbourg, France*
[3] *French Alternative Energies and Atomic Energy Commission,*
*Bruyères-le-Châtel, France*
[4] *Seismology at School in Nepal, Pokhara, Nepal*
[5] *Earthquake and Geophysics Division, Department of Geology and Mines,*
*Thimphu, Bhutan*

## 4.1. Introduction

Seismology is a powerful geophysical method to image structures and ruptures in the Earth's interior at various spatial scales. In the Himalaya and Tibetan Plateau system, both active-source and passive-source seismic experiments have imaged the physical properties of the underground and its geological structures. These images reveal the inner architecture of this collision zone from shallow to large depths, which reflect the result of millions

*Himalaya, Dynamics of a Giant 1,*
coordinated by Rodolphe CATTIN and Jean-Luc EPARD.
© ISTE Ltd 2023.

of years of orogenic processes and their interactions. At the same time, these field experiments record numerous earthquakes that highlight the currently active geological structures and thereby the ongoing brittle deformation pattern of the lithosphere.

Seismological exploration of the crust and the lithosphere in the region has been carried out first perpendicularly to the mountain belt, to characterize the main trends in its structure. Further experiments followed with time, shedding light on along-arc variations as well. Instrumental seismicity is detected by stations around the globe for large magnitude events, and locally deployed seismometer networks allowed to delineate features that cause moderate and very small earthquakes. However, due to the very large extent of the area as well as to challenges in accessing key areas, the true three-dimensional structure and deformation of this system, in terms of discontinuity geometries, wave-speed anomalies and seismic activity, is still far from being complete and homogeneously mapped. This chapter presents the main features hitherto imaged and detected by seismology.

## 4.2. Imaging by elastic waves

Imaging using seismic waves is the only method with which we can explore every part of our planet at depth. This is possible due to two aspects of nature. First, seismic waves are relatively little attenuated with distance; beyond the geometrical spreading of energy, they are efficiently propagating through rocks in the underground. This efficiency depends on the wave mode and the wave's frequency, which enables a versatile use of seismic waves in imaging. Second, earthquakes liberate energy that is focused in time and space, and ranges from very small to very large amounts, thus providing an outstanding source of seismic waves.

Humans have attempted to create alternative sources in a controlled manner, and succeeded so for a limited range of energies. Beyond the generation and the propagation of seismic waves, their observations enable exploration of the underground through seismic imaging. The quality and resolution of these images depend primarily on the depth of the target structure, the frequency of the source generating the waves, and the type and geometry of sensors detecting the wave field.

## 4.2.1. *Active seismics*

In active-source seismic imaging (Figure 4.1a), the energy is introduced to the ground by a manmade device. A hammer shot on a metal plate or a repeatable weight-drop system are good point sources at the surface (see Volume 1 – Chapter 7). At depth, and where permits allow, explosives can be used. For a controlled range of input frequencies, vibrating trucks can be employed. Other solutions, such as an air gun in water, may provide a suitable source. The depth penetration of such campaigns reaches between meter to tens of kilometer, while the horizontal propagation typically reaches 10 times more; therefore, usually a Cartesian coordinate system is used for simulating how waves propagate. Most often, P (primary, compressional) waves are used, and the direct, the reflected and refracted phases are analyzed. In some cases, S (secondary, shear) waves, multiples and surface waves can also be exploited. Detection is most often performed by geophones, usually one (vertical) component, relatively higher frequency sensors measuring ground motion.

## 4.2.2. *Passive seismics*

In passive-source seismological imaging (Figure 4.1b), the energy is from a natural source, and the classical and best-focused sources are earthquakes. Depending on their magnitude, the generated waves can travel very large distances, easily across or around the Earth, leading to the use of a spherical coordinate system for calculating ray paths.

P and S waves are emitted naturally, and their interaction with the surface generates surface waves. All of these, and their combination, for example, P-to-S converted waves, are exploited to image Earth structure. Since recently, not only earthquake signals, but also the background noise from the environment (mostly generated by the oceans) was successfully recorded and used to enable imaging within a network of seismometers. These usually have three components to record ground motion, and are sensitive to a larger range of frequencies. Many other natural sources such as volcanoes, meteorite impacts and landslides can be detected, but are rarely used for imaging Earth's interior.

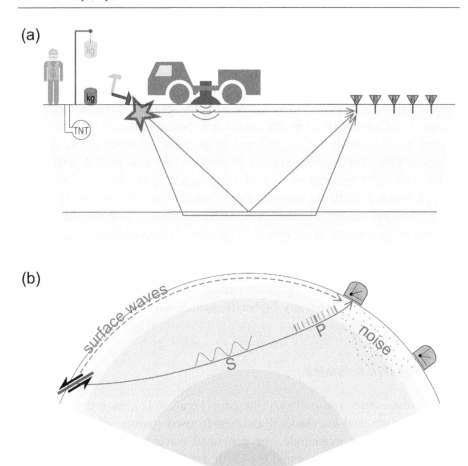

**Figure 4.1.** *Main elements of seismic imaging, from source to receivers. (a) Active-source imaging involves human action to trigger a source at a known location and time, such as an explosive, a repeatable weight-drop system, a hammer shot, a vibrating truck or other ingenious solution. The generated direct, reflected and refracted waves are recorded by geophones at relatively shorter distances. (b) Passive-source imaging uses seismic waves generated by a natural source, typically earthquakes. Various waves (such as P, S and surface waves) are recorded at large distances. More recently, the ambient environmental noise proved to be useful in imaging a region beneath a network of stations. For a color version of this figure, see www.iste.co.uk/cattin/himalaya1.zip*

### 4.2.3. Tomographic imaging for bulk properties

Although the word "tomography" literally means "description by sections", in seismology, it is mostly understood that it refers to cross-sections or depth-sections on which seismic wave-speed anomalies are shown. These are calculated by measuring the travel times of a given set of seismic waves, and then solving a mathematical inverse problem to locate anomalies: bulk volumes where waves travel slower or faster than an initially expected average. This method works well if numerous stations detect waves from numerous sources, and the ray-paths connecting these two sets of points cross each other as much as possible. The resulting seismic anomalies represent smaller or larger volumes of rocks whose seismic wave-speed stands out due to their temperature, composition, fluid-content, anisotropic property or some other reasons. Tomographic studies are most often carried out using P-waves thanks to the relative easiness to pick their arrival on seismograms; however, S-wave, surface-wave and ambient noise tomography studies are more and more frequent. While the overall slower or faster anomalies are undoubtedly present in well-resolved areas of the tomographic images, the edges of these anomalies are rarely imaged in a sharp way.

### 4.2.4. *Wave reflections and conversions for interfaces*

Sharp boundaries – also called interfaces or discontinuities – are much better resolved using waves that are either reflected at such a boundary or transmitted and mode-converted when crossing such a boundary.

*Reflection* is very often used in active seismics where both the source and the receivers are *above* the interface. Farther away from the source, such rays do not reflect any more, but penetrate the layer below and propagate along the interface: this is a *refracted* wave that then sends energy back to the surface (Figure 4.1a).

However, when a passive-source wave impinges at a discontinuity *from below*, it is often its mode-conversion property that is exploited. For example, a small part of the P wave's energy arriving below a discontinuity is converted to generate an S-wave above it. This happens because the arriving P wave comes at an oblique angle to the discontinuity – if it arrives perpendicularly, no energy is converted.

All these approaches are useful in constraining the location of discontinuities, sharp velocity changes at depth, and thus to draw the structural contours of the underground.

## 4.3. Exploring the Central Himalaya along cross-sections

### 4.3.1. *Field experiments*

Many fields seismological experiments have been performed over the last four decades (Figure 4.2).

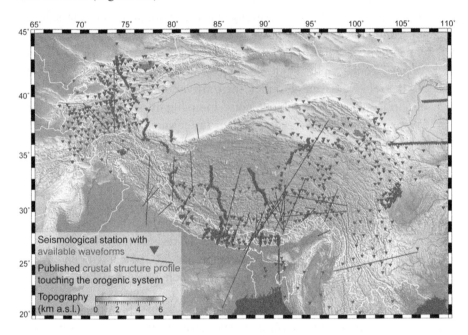

**Figure 4.2.** *Map of seismological data coverage of the Himalaya–Tibet system. Red symbols represent a minimum coverage: further stations exist but their data is not openly available. Brown lines show the location of relevant cross-sections touching the orogenic system. For a color version of this figure, see www.iste.co.uk/cattin1/himalaya1.zip*

The deep exploration of the Himalaya–Tibet orogenic system started with active-source studies; however, by now, there have been more passive-source data collected. Unfortunately, not all of these are available for research: while temporary seismic networks installed for short times (approximately 0.5–2 years) have resulted in publicly available datasets, raw data from permanent observatory sites are, for the majority, not accessible. Therefore, our knowledge of the area relies on open but short-duration datasets, and published cross-sections and maps that represent data in much reduced ways (stacked, projected, heavily processed, interpreted). Nevertheless, it is generally considered that the main features of the Himalayan structure at depth are now known.

## 4.3.2. Main interfaces

It is obvious that the crustal structure beneath the approximately 5-km high topography of the Tibetan Plateau must be different from the classical approximately 35-km thick architecture that prevails in the Indian peninsula. What happens to this crust as it enters the Himalayan collision zone? Already in the 19th century, well before Mohorovićič pointed out the existence of the crust–mantle boundary (which was later coined the "Moho"), it was known to Earth scientists from gravimetric data (see Volume 1 – Chapter 5) that the Himalayan mountains must have a root: the lighter crust must be thicker, at the expense of the denser mantle. But how thick is the crust beneath Tibet? And how does this thickening occur? These are the main questions to which seismology could give a clear answer by imaging the main interfaces.

### 4.3.2.1. The Main Himalayan Thrust

The Main Himalayan Thrust (MHT) deserved its name as it is along this fault that the downgoing India plate slides beneath the Himalayan orogen. It is the equivalent of a subduction zone megathrust, and the very fault on which the major Himalayan earthquakes occur. The inferences about the MHT geometry came partly from geology and different surface expressions of this fault in the past (see MCT, MBT, MFT and Volume 1 – Chapter 3), partly from local geophysical data in Central Nepal (e.g. magnetotellurics data, Lemonnier et al. 1999), and, last but not least, active seismics. Indeed, it is in the major geophysical effort of the INDEPTH project (e.g. Nelson et al. 1996; Hauck et al. 1998, Figure 4.3) that the steep part of the MHT was imaged for the first time: a prominent reflector beneath the Himalaya, dipping North by locally as much as 15°. This key experiment took place in the Yadong–Gulu rift, where the Himalayan 8,000 m peaks cease and access with geophysical equipment was possible. Due to this strategic location, the link between south and north of the Himalaya was established.

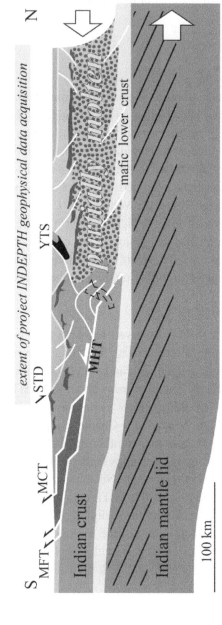

**Figure 4.3.** Interpretative cross-section of the INDEPTH experiment results, a composite picture drawn from active and passive seismology, magnetotellurics and geology data. Redrawn and modified from Nelson et al. (1996). For a color version of this figure, see www.iste.co.uk/cattin/himalaya1.zip

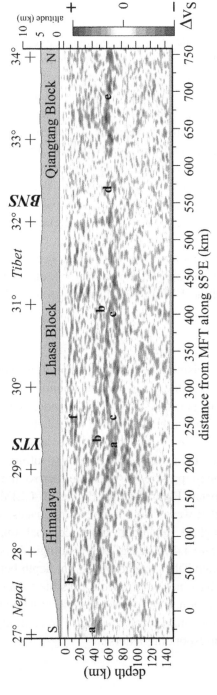

**Figure 4.4.** *Migrated P-to-S converted waves' cross-section along 85°E from project Hi-CLIMB. $\Delta V_S$ is the shear-wave velocity change with depth. The letters mark features explained in the text. YTS: Yarlung–Tsangpo Suture. BNS: Banggong–Nujiang Suture. Image from Hetényi (2007), then Nábělek et al. (2009). For a color version of this figure, see www.iste.co.uk/cattin/himalaya1.zip*

Two zones with fluids in the crust were identified early in the exploration of this area. One of them in Nepal, associated with the MHT where it is still flat at approximately 12 km depth. Based on magnetotellurics sounding results, and subsequent seismological experiments which detected low seismic velocities at that depth, it has been proposed that aqueous fluids accumulated at the MHT shear zone cause this anomaly, and this interpretation is still generally accepted. The other zone with fluids was proposed by the INDEPTH experiment, in southern Tibet, north of the point to which the MHT could be imaged. There, at a shallower depth of approximately 15–20 km, strong reflections in active seismics and a velocity-decrease seen in passive seismics, together with high electrical conductivity models, have been interpreted as partial melt in the crust. Although only the top of such a partially molten zone could be well detected (and the bottom not), the interpretation of the results went until proposing that more than half of the Tibetan crust was partially molten (Figure 4.3). Although this inference has proven not to be ubiquitous across the Tibetan Plateau, it has opened the room for a class of numerical models in which the middle of the Tibetan crust flows at geologically fast rates (see discussion below).

### 4.3.2.2. *The Moho*

Both active and passive seismological data acquired as early as the 1980s and 1990s have hinted at the crust–mantle boundary being locally at approximately 80 km depth beneath Tibet. This is more than double of the normal crustal thickness. Therefore, the search for a better image of the geometry, and for a better understanding of the processes able to create such structures, has continued since then. Serious field efforts ultimately yielded the spatially densest and longest passive-seismic line, across the Himalaya and to the centre of the Tibetan Plateau at 85°E: the Hi-CLIMB project. With the use of P-to-S converted waves from remote earthquakes, this project has produced the hitherto highest resolution cross-sectional image of the Himalayan collision zone (Figure 4.4). The earlier seismological results hinted at the depth of the Moho beneath Tibet around 60–80 km depth below the sea level. The Hi-CLIMB results have not only made it possible to precisely map this value along a long cross-section, but have succeeded in imaging the Moho of the India plate in a continuous manner all the way. The resulting image in Figure 4.4 shows (with the corresponding letters):

(a) the clear Moho beneath the Ganges Basin and Nepal, and a smooth deepening from 40 km beneath India to approximately 73 km at the Yarlung–Tsangpo Suture;

(b) the Main Himalayan Thrust observed continuously between Nepal, along a ramp, extending until 450 km north of the orogenic front;

(c) the Indian lower crust between the two previous interfaces, underlies most of southern Tibet (the Lhasa Block);

(d) a relatively weaker Moho signal beneath the Banggong–Nujiang Suture;

(e) coherent Moho signature at approximately 65 km depth beneath northern Tibet (Qiangtang Block);

(f) some of the shallow crustal "bright spots" (partial melt) in southern Tibet.

### 4.3.2.3. *Interpretation: underthrusting*

The primary result when interpreting this image is that the India plate underthrusts southern Tibet, in a surprisingly horizontal way, over a distance of about 200 km once it has reached its maximal depth (Figure 4.5). This underthrusting has a series of implications: (1) the India/Tibet boundary at depth is not at the same position as geologically inferred at the surface, at the Yarlung–Tsangpo Suture. There is decoupling between the various geological layers at depth; (2) the crustal thickness of southern Tibet, considering 73 km Moho depth below sea level and 5 km topography, is approximately 78 km, and it is approximately 70 km in northern Tibet although the elevations are the same. This setting strongly conditions the hotter temperature field in the Tibetan crust. The difference in Moho depth despite the same topography can be reconciled with a denser, eclogitized Indian lower crust; (3) the fate of the India plate is resolved: it currently extends over 450 km beneath the Himalaya–Tibet system, and then plunges nearly vertically (e.g. Tilmann et al. 2003); (4) opposite to the northward propagating India plate, the southward propagating deep portion of the Eurasia plate reaches until around the Banggong–Nujiang Suture, these motion directions are expressed by dipping seismic fabrics at depth.

These interpretations still hold today for the structure of the collision zone at its central longitudes, and fit well with the conclusions drawn from gravity anomalies and related modeling (see Volume 1 – Chapter 5).

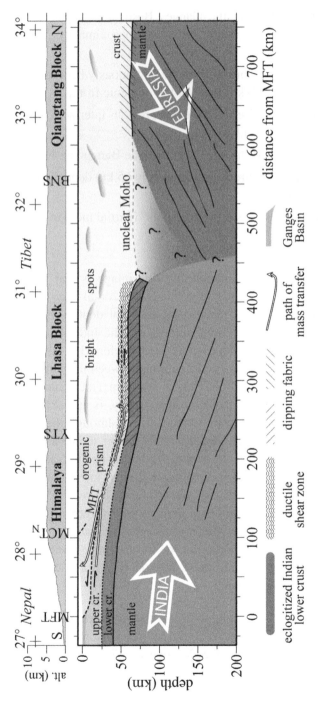

**Figure 4.5.** *Synoptic cross-section of the Himalaya–Tibet collision zone at its central longitudes, based on project Hi-CLIMB results along 85°E. Image from Hetényi (2007), later published in Nábělek et al. (2009). For a color version of this figure, see www.iste.co.uk/cattin/himalaya1.zip*

### 4.3.2.4. *Melt in the Tibetan crust?*

When the INDEPTH results proposed a large amount of partial melt in the Tibetan crust (Figure 4.3), a new class of so-called "channel-flow models" were developed (see Volume 1 – Chapter 3). These numerical models simulated hot crust to fit the geological record, primarily from the Miocene, to propose the presence of quickly deforming crustal material in Tibet. Serious debate followed around these models, in which it was – and it still is – often forgotten that geophysical data shows the current structure while the rock record reflects the geological past, and that widespread partial melting may have well existed when Himalayan leucogranites formed but is unlikely to underlie most of the plateau today.

Results from the Hi-CLIMB data have brought important constraints in this debate, by significantly limiting the volume of partial melt compared to INDEPTH. Mapping a much larger area than INDEPTH, and by employing passive-source imaging that illuminates the entire crust from below, Hi-CLIMB results have shown that partial melt is located in relatively thin, approximately 10 km thick pockets that do not seem to exceed 50 km in length and are not interconnected (Hetényi et al. 2011). This spot-wise, localized presence of partial melt has then been corroborated by magnetotellurics surveys (e.g. Rippe and Unsworth 2010; Wei et al. 2010; Xie et al. 2016), which precludes channel-flow models being a viable mechanism of Tibet's current deformation.

Finally, the geotherm in northern Tibet seem to be higher than in southern Tibet, as reflected by poor propagation (quicker attenuation) of refracted P- and S-waves. This can easily be explained by the absence of a cold, underthrusting mantle lithosphere as it is the case of the India plate beneath southern Tibet, but does not necessarily lead to widespread partial melting in the north either, according to results available so far.

### 4.3.3. *Where do subducted plates go?*

As mentioned above, the India and Eurasia plates meet at a depth just south of the Banggong–Nujiang Suture. Results from the follow-up of INDEPTH experiments have imaged with tomography that India then plunges nearly vertically to approximately 350–400 km depth (see Tilmann et al. 2003). The fate of the currently underthrusting India plate then seems to be clear, but this is only part of its history. There have been at least one, but likely more detachments of the India plate subducting since approximately 50 Ma (e.g. Guillot et al. 2003), and these fragments have sunken further down. Global passive-source tomography has identified these fragments, and they currently lie at great depths (1,000–2,300 km) beneath the Indian peninsula, as, since their subduction, they have been overridden by the northward shifting India plate (e.g. van der Voo et al. 1999; Replumaz et al. 2004).

Such subduction, detachment and underthrusting scenarios have a primary influence on the long-term dynamics of the upper mantle (see Volume 1 – Chapter 2). They influence the evolution of the temperature field, and also the flow direction in the asthenosphere as well as the evolution of the mantle transition zone. Due to the complex history and large area at hand, only parts of this evolution have been reconstructed, and there are still open questions for which 3D imaging and modeling efforts are required to obtain an answer.

## 4.4. Lateral variations

The picture drawn above is of the main features that characterize the Central Himalaya. There are variations at smaller and at larger spatial scales, both of which are only partially imaged and understood so far. This section describes three prominent elements.

### 4.4.1. *Lateral ramps on the MHT, along-arc Moho variations*

Seismic experiments with spatially dense station spacing provide high-resolution images of the crust and the lithosphere. As times passed, more and more of these experiments took place in the Himalaya (see Figure 4.2). They had diverse length and resolution power; nevertheless, the majority of these experiments have succeeded in imaging the two main discontinuities: the MHT and the Moho.

The direct comparison of these structures between profiles along the Himalayan arc reveals that the architecture of the orogen is not cylindrical. Even within the Central Himalaya, variations in depth and shape of the MHT and the Moho exist (Figure 4.6). Although some of the differences may be attributed to different data processing methods and velocity models used to convert time observations to depth interpretations, a good part of the variations are real. This is not too surprising considering the width of the area covered by these experiments, which is approximately 1,500 km, and the geological variability both at the surface and at depth. It is known that the topography changes along the arc, for example, that the peaks exceeding 8,000 m elevation are not everywhere. The width of different lithological units outcropping in the Himalaya also varies in an east-west direction (see Volume 2). Microseismicity studies as well as seismic and geological studies published in the aftermath of the 2015 Gorkha earthquake (Volume 3 – Chapter 8) reveal that the Main Himalayan Thrust's shape has lateral ramps, and that even a kilometer-scale depth change between adjacent segments can affect rupture propagation during earthquakes. Finally, the inherited structure of the downgoing India plate (see Volume 1 – Chapter 1) has a direct impact on the Moho geometry and its lateral variation (see Figure 4.6).

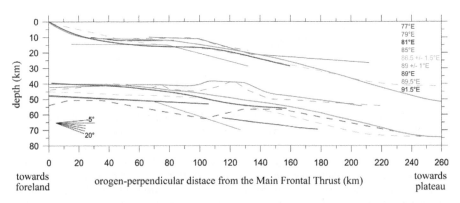

**Figure 4.6.** *Compilation of Moho and MHT geometry across the Himalaya at various longitudes shown in different colors. Figure modified from Subedi et al. (2018). For a color version of this figure, see www.iste.co.uk/cattin/himalaya1.zip*

### 4.4.2. *Segmentation of the India plate lithosphere*

Beyond indicators of segmentation south of and in the Himalayan arc, recent seismic tomography focusing on refracted waves – which in this case propagate just under the Moho – revealed distinct NNE-SSW-oriented low-velocity stripes in a high-velocity background beneath Southern Tibet (Li and Song 2018). The authors interpret this observation as fractures that separate mantle lithosphere segments (Figure 4.7). Their origin could be inherited, which means that this inheritance was preserved through millions of years of underthrusting and related deformation while passing beneath the

Himalaya. Their fate, or further descent of these segments at depth, may also be independent of each other. Most importantly, the fracture or tear zones between them would allow upwelling from the hot asthenosphere, inducing partial melting in the crust and contributing to rifting at the surface.

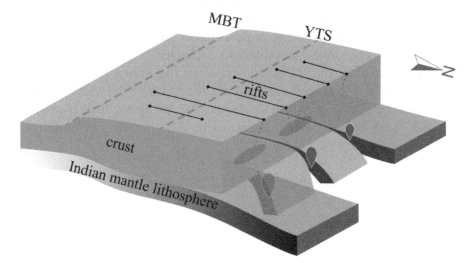

**Figure 4.7.** *Schematic cartoon on how the Indian mantle lithosphere may be torn beneath southern-central Tibet and how it may affect the crust. The tears in the mantle lithosphere are based on seismic tomography results focusing on that depth range. Tears may induce local upwelling and/or temperature increase, and trigger partial melting in the crust, ultimately causing extension and rifts at the surface. However, a one-to-one relationship between tears, molten areas and rifts could not (yet) be established. The figure is not to scale, and is redrawn after Li and Song (2018). For a color version of this figure, see www.iste.co.uk/cattin/himalaya1.zip*

While the qualitative concept is appealing, the authors note that rifts are not located directly above these fractures zones. Together with the description of localized partial melt in the Tibetan crust (see section 4.3.2.4), we can safely claim that these recent results cannot be confirmed yet. Higher resolution 3D imaging and a quantitative assessment of melt percentage and melt migration would be needed to do so.

Nevertheless, inheritance certainly plays a role in shaping the roots of the Himalayan orogen, as shown, for example, by gravimetric investigations in Volume 1 – Chapter 5. Also, the seismically active Dhubri–Chungthang Fault zone crossing the Himalaya at depth at the longitude of Sikkim (Diehl et al. 2017; Michailos et al. 2021) attests to the actuality of inheritance. For an overview of short- and long-term indicators of lateral segmentation of the Himalaya, we refer to Dal Zilio et al. (2021), especially their Figure 5.

### 4.4.3. *The western and eastern syntaxes*

Zooming out to the entire Himalayan orogen, large-scale open questions still persist. In this chapter so far, the Central Himalayan picture over 1,500 km east-west distance extent was presented, with its main feature being sub-horizontal underthrusting, followed by near-vertical subduction.

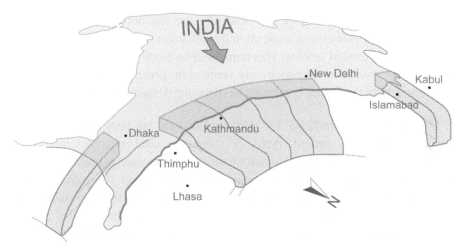

**Figure 4.8.** *Present-day geometry of the subducting Indian plate that differs beneath the Himalayan arc (underthrusting) and the surrounding orogens (subduction), as inferred from geophysical explorations. Orange lines indicate the frontal thrust faults, the Main Frontal Thrust is shown by the thicker line. The large arrow is the present-day convergence direction of the India plate with respect to Eurasia. Figure modified from Dal Zilio et al. (2021). For a color version of this figure, see www.iste.co.uk/cattin/himalaya1.zip*

However, when we focus on the two syntaxes, the collision geometry at depth is different. In the east, seismicity and structural seismology shows that the plate beneath the Indian peninsula subducts beneath the Indo-Burman ranges, towards to the east (Figure 4.8; e.g. Bora et al. 2022). In the west, intermediate-depth seismicity and seismic tomography reveals a very complex interaction between the downgoing slab beneath the Hindu-Kush (Figure 4.8; Kufner et al. 2016), against which a piece of Eurasian lithosphere is subducted beneath the Pamir towards the south. For the India plate, this means that there must be a significant change from the collision and deformation style when moving away from the central longitudes. Both towards the east and towards the west, several hundreds of kilometers along the Himalaya remain without a reliable, high-resolution image that would reveal how such changes in large-scale geodynamic processes occur.

## 4.5. Current seismicity of the Himalaya

### 4.5.1. *Earthquake detection, location and activity*

Elastic waves recorded by seismometers are not only used for seismic imaging: their primary use is most often the detection of earthquakes and other signals causing ground motion. This domain of research and monitoring duties are an art on their own, of which only some main principles are summarized here. For details the reader is referred to specialized manuals and books.

Ground motions induced by earthquakes can be very large to very small. Accordingly, various types of sensors, called seismometers, can be used to detect the signals: accelerometers for strong motions, short period velocity-meters for nearby earthquakes and broad (frequency)-band velocity-meters for faraway earthquakes. The sensitivity of the instrument, the quietness (low environmental noise level) of the site and the setup of how a sensor is installed are key elements to successful and clean recordings.

Nowadays, seismometers operate continuously, and it is usually computers that detect the arrival of elastic waves. Still, human verification is useful to confirm not only the correctness of detection, but also that nearly simultaneous detections at several stations correspond to a real earthquake. The precise location of an earthquake requires several conditions to be met, among which a good local wave-speed model, consistent user experience, as well as advanced and robust techniques can be mentioned. Nevertheless, the most important criteria are probably the seismic network configuration, the geometry of deployed seismometers.

If an earthquake happens within a group of four or more seismometers, the location is by default better quality than if it happens outside that zone. To avoid any luck factor, seismic network geometries can therefore evolve with time to better cover a given group of earthquakes. Since small magnitude earthquakes release small amounts of energy, and because the generated elastic waves' amplitude decreases in three dimensions of the surrounding rock volume, smaller events can only be located if there is a sufficient number of seismometers nearby. Overall, the stations' spatial density and geometry are the key factors in mapping seismicity before data processing starts.

Finally, and luckily for humanity, nature has a beautiful scaling law about earthquake size: larger magnitude events are rarer than smaller magnitude events. The beauty lies in the numbers; statistically, for a given region and a given time period, there are 10 times more magnitude M-1 events than magnitude M events. For example, there is, on average, one M8 earthquake per year on Earth, then there are 10 M7, 100 M6, 1,000 M5 events and so on. These numbers are approximate, and it should not be expected that *every* year produces these numbers.

While larger magnitude events on Earth can be detected from the existing seismometers around the globe, located mostly on continents and islands, smaller magnitudes cannot and this requires local seismic networks to be installed. This represents a major effort in the field to find suitable sites and install seismometers, but already with one or few years of operation hundreds and thousands of local, M3, M2 and even M1 earthquakes can be located. This is the only way to see where active faults are, how active they are, and how deformation of the crust and the lithosphere is accommodated on the geologically shortest timescales.

### 4.5.2. *Seismicity of the Himalaya: an incomplete patchwork*

A compilation of earthquakes in the Himalaya detected by seismometers and available from nine, high-quality open catalogues is shown in Figure 4.9 (sources: Monsalve et al. 2006; Adhikari et al. 2015; Baillard et al. 2017; Diehl et al. 2017; Hoste-Colomer et al. 2018; USGS 2022; NEMRC 2022; Laporte et al. 2021; Michailos et al. 2021). The input data is very heterogeneous; therefore, it has to be read carefully. Three major types of data can be distinguished as follows:

First: moderate to large earthquakes, which have been detected by seismometers around the world. The elastic waves traveled far enough to obtain a picture of such events in the entire region, down to typically M5 or even M4 earthquakes. These data are available for a few decades only.

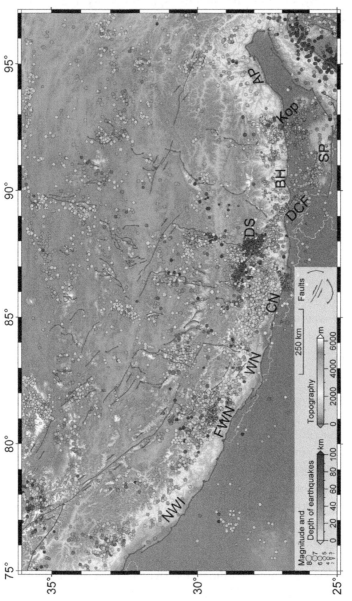

**Figure 4.9.** *A compilation of earthquakes from multiple, high-quality open catalogues in the Himalaya–Tibet system. Earthquakes are coloured according to their depth, and circle size is proportional to magnitude. Faults are from Styron et al. (2010), boundaries are shown in white for reference. See the text for the interpretation of this map. For a color version of this figure, see www.iste.co.uk/cattin/himalaya1.zip*

Second: small earthquakes, which have been detected by temporary seismic networks. In the Himalaya, such networks have been recording signals locally between six months and three years at a time. Because of the location principles mentioned in section 4.5.1, only earthquake loci within the respective seismic networks can be considered reliable, providing zones of well-resolved, low magnitude (M3 and below) events.

Third: the absence of earthquakes. The map shows several areas without any earthquakes. The lack of small events is normal, as most regions have not been covered by a local seismic network. The lack of large events is also normal, because a few decades of globally detected earthquake data is much shorter than the duration of the seismic cycle. The seismic cycle is the period of time between two large events in an area. This can be between tens and thousands years long, and in the Himalaya, it is considered to be between few to several hundreds of years (see Volume 3 – Chapter 5 on historical earthquakes and paleoseismicity). As a result, "no data" on the map has to be read carefully: is it a true absence of earthquakes? Or simply absence of seismic activity since instrumental detection started?

### 4.5.3. *Seismicity of the Himalaya: main features*

Figure 4.9 shows seismicity as well as main tectonic faults mapped at the surface. It can be seen that a large portion of these faults are seismically not active in the geologically youngest time period. Some of the north-south trending grabens in southern Tibet are active, with earthquake depths usually in the upper crust, and in some places deeper.

Surprisingly, the Main Frontal Thrust (MFT) at the southern edge of the Himalaya does not appear to be seismically active. As described in Volume 1 – Chapter 3, this fault becomes sub-horizontal and penetrates beneath the Himalaya. It is the megathrust along which the India plate goes beneath the Tibetan Plateau. Seismicity on this fault, the Main Himalayan Thrust, appears approximately 100 km further north of the MFT, and can be well followed along the Himalayan belt. In particular, in Central Nepal (CN), the aftershocks of the 2015 Gorkha earthquake show very well the extent of that fault zone (Adhikari et al. 2015; Baillard et al. 2017, see Volume 3 – Chapter 8). Similarly, in Far Western Nepal (FWN), a local seismic network has detected continuous activity on and around the MHT (Hoste-Colomer et al. 2018; Laporte et al. 2021). In Western Nepal (WN) however, seismic activity seems to be lower even from a global perspective, which is worrisome considering that there has been no major earthquake there since 1505.

In the Indian parts of this belt, there is less high-quality data available, but both North-Western India (NWI) and the north-eastern territory of Arunachal Pradesh (AP) show activity on the MHT. Both zones have ruptured in major earthquakes in the past centuries.

There is an intriguing difference between the western and central part of the Himalaya, and its eastern third: in the former segment, the foreland basin is wide and deep, while in the latter, the Shillong Plateau (SP) is a locally high relief feature with almost no sediments in the foreland. This latter zone is seismically active, unlike the foreland in the central and western Himalaya (see also the 1897 earthquake in Volume 3 – Chapter 5). This means that the eastern Himalayan foreland and the Shillong Plateau is a zone of diffuse deformation, with earthquakes occurring in a larger area (e.g. Grujic et al. 2018). In its eastern part, the Kopili fault zone (Kop) broadly connects the Shillong Plateau and goes slightly beneath the Himalaya. In the western part, the Dhubri–Chungthang Fault (DCF) zone is a very linear, narrow, mid-crustal zone of seismicity, which can be traced beneath the MHT and the Himalaya (Diehl et al. 2017). Between the two, the Bhutan Himalaya (BH) is apparently less seismically active in the instrumental observation time, although it has also produced large earthquakes in the past.

The DCF is a major transition between central and eastern Himalaya, also detected in gravity anomalies, and is most likely a geologically inherited feature. The plate east of this line may be a different terrain than the India plate. Furthermore, the DCF can be tracked further down to the root of the Himalaya, where deep seismicity (DS) at 60–80 km depth occurs (Monsalve et al. 2006) along an at least 200 km long, nearly east-west segment (Michailos et al. 2021). Whether this is a tectonic tear in the plate or a dehydration front during eclogitization of the lower crust, or both, is still a question, the same as what proportion of these earthquakes may occur in the mantle and the lower crust (Schulte-Pelkum et al. 2019; Michailos et al. 2021).

## 4.6. Conclusion

Seismological imaging has revealed the most important part of the subsurface structure of the Himalayan orogen, and also enabled other geophysical methods to interpret their datasets in terms of deep structure. As of today, numerous structural profiles crossing the Himalaya are published and some of the raw data is also available. These results have brought the community further in the past 20 years from a generally cylindrical view of the Himalaya, to a perception where small- and large-scale lateral variations play an important role.

The picture of current Himalayan seismic activity is inherently incomplete, but time is in the favur of our understanding. Global networks detect all moderate and large events, and further temporary networks zoom into the key parts of the orogen. Such glimpses into seismic activity are extremely useful, even though they are not representative of the seismic cycle. The current pattern reflects well the main structural features of the orogen, namely, underthrusting, lateral variations and even some key across-strike features such as the localized Dhubri–Chungthang Fault zone.

Nevertheless, our images of structures and of seismicity – and therefore our understanding of processes – are still incomplete. The step from 2D images towards 3D reality is often more complex than we may think, especially because the scale of variations cannot be always properly apprehended. The seismicity puzzle is being slowly put together both in time and space, but the whole is still far from being well understood. Further work is required, with multiple objectives. Technically, to deploy spatially denser seismological arrays, complemented by some large-aperture arrays, with some stations staying for longer times. Politically, to convince the authorities about the need for science to benefit from accessing particular geographical areas as well as from open data policies. Methodologically, to simultaneously include small- and large-scale information, sharp changes and gradients, and to jointly use datasets from several methods. Finally, disciplinarily, to better link across various geoscience domains, as only such approaches will resolve the inherent spatial and temporal variabilities of natural processes, such as, for example, the role of million (billion?) year-long geological inheritance on the lateral extent of seconds-long earthquake rupture.

## 4.7. References

Adhikari, L.B., Gautam, U.P., Koirala, B., Bhattarai, M., Kandel, T., Gupta, R.M., Timsina, C., Maharjan, N., Maharjan K., Dahal, T. et al. (2015). The aftershock sequence of the April 25 2015 Gorkha-Nepal earthquake. *Geophys. J. Int.*, 203(3), 2119–2124.

Baillard, C., Lyon-Caen, H., Bollinger, L., Rietbrock, A., Letort, J., Adhikari, L.B. (2017). Automatic analysis of the Gorkha earthquake aftershock sequence: Evidences of structurally segmented seismicity. *Geophys. J. Int.*, 209(2), 1111–1125.

Bora, D.K., Singh, A.P., Borah, K., Anand, A., Biswas, R., Mishra, O.P. (2022). Crustal structure beneath the Indo-Burma Ranges from the teleseismic receiver function and its implications for dehydration of the subducting Indian slab. *Pure Appl. Geophys.*, 179, 197–216.

Dal Zilio, L., Hetényi, G., Hubbard, J., Bollinger, L. (2021). Building the Himalaya from tectonic to earthquake scales. *Nat. Rev. Earth Environ.*, 2, 251–268.

Diehl, T., Singer, J., Hetényi, G., Grujic, D., Giardini, D., Clinton, J., Kissling, E., GANSSER Working Group (2017). Seismotectonics of Bhutan: Evidence for segmentation of the Eastern Himalayas and link to foreland deformation. *Earth Planet. Sci. Lett.*, 471, 54–64.

Grujic, D., Hetényi, G., Cattin, R., Baruah, S., Benoit, A., Drukpa, D., Saric, A. (2018). Stress transfer and connectivity between the Bhutan Himalaya and the Shillong Plateau. *Tectonophysics*, 744, 322–332.

Guillot, S., Garzanti, E., Baratoux, D., Marquer, D., Mahéo, G., de Sigoyer, J. (2003). Reconstructing the total shortening history of the NW Himalaya. *Geochem. Geophys. Geosys.*, 4, 1064.

Hauck, M.L., Nelson, K.D., Brown, L.D., Zhao, W., Ross, A.R. (1998). Crustal structure of the Himalayan Orogen at approximately 90 degrees east longitude from Project INDEPTH deep reflection profiles. *Tectonics*, 17(4), 481–500.

Hetényi, G. (2007). Evolution of deformation of the Himalayan prism: From imaging to modelling. PhD Thesis, École Normale Supérieure, Paris.

Hetényi, G., Vergne, J., Bollinger, L., Cattin, R. (2011). Discontinuous low-velocity zone in southern Tibet questions the viability of channel flow model. In *Growth and Collapse of the Tibetan Plateau*, Gloaguen, R. and Ratschbacher, L. (eds.). Geol. Soc. London Spec. Pub., London.

Hoste-Colomer, R., Bollinger, L., Lyon-Caen, H., Adhikari, L.B., Baillard, C., Benoit, A., Bhattarai, M., Gupta, R.M., Jacques, E., Kandel, T. et al. (2018). Lateral variations of the midcrustal seismicity in western Nepal: Seismotectonic implications. *Earth Planet. Sci. Lett.*, 504, 115–125.

Kufner, S.K., Schurr, B., Sippl, C., Yuan, X., Ratschbacher, L., Akbar, A.M., Ischuk, A., Murodkulov, S., Schneider, F., Mechie, J. et al. (2016). Deep India meets deep Asia: Lithospheric indentation, delamination and break-off under Pamir and Hindu Kush (Central Asia). *Earth Planet. Sci. Lett.*, 435, 171–184.

Laporte, M., Bollinger, L., Lyon-Caen, H., Hoste-Colomer, R., Duverger, C., Letort, J., Riesner, M., Koirala, B.P., Bhattarai, M., Kandel, T. et al. (2021). Seismicity in far western Nepal reveals flats and ramps along the Main Himalayan Thrust. *Geophys. J. Int.*, 226(3), 1747–1763.

Lemonnier, C., Marquis, G., Perrier, F., Avouac, J.P., Chitrakar, G., Kafle, B., Sapkota, S., Gautam, U., Tiwari, D., Bano, M. (1999). Electrical structure of the Himalaya of Central Nepal: High conductivity around the mid-crustal ramp along the MHT. *Geophys. Res. Lett.*, 26(21), 3261–3264.

Li, J. and Song, X. (2018). Tearing of Indian mantle lithosphere from high-resolution seismic images and its implications for lithosphere coupling in southern Tibet. *Proc. Natl. Acad. Sci. USA*, 115(33), 8296–8300.

Michailos, K., Carpenter, S.N., Hetényi, G. (2021). Spatio-temporal evolution of intermediate-depth seismicity beneath the Himalayas: Implications for metamorphism and tectonics. *Front. Earth Sci.*, 9, 742700.

Monsalve, G., Sheehan, A., Schulte-Pelkum, V., Rajaure, S., Pandey, M.R., Wu, F. (2006). Seismicity and one-dimensional velocity structure of the Himalayan Collision Zone: Earthquakes in the crust and upper mantle. *J. Geophys. Res. Solid Earth*, 111, 1–19.

Nábělek, J., Hetényi, G., Vergne, J., Sapkota, S., Kafle, B., Jiang, M., Su, H., Chen, J., Huang, B.S., Hi-CLIMB Team (2009). Underplating in the Himalaya-Tibet collision zone revealed by the Hi-CLIMB experiment. *Science*, 325(5946), 1371–1374.

Nelson, K.D., Zhao, W.J., Brown, L.D., Kuo, J., Che, J.K., Liu, X.W., Klemperer, S.L., Makovsky, Y., Meissner, R., Mechie, J. et al. (1996). Partially molten middle crust beneath southern Tibet: Synthesis of project INDEPTH results. *Science*, 274(5293), 1684–1688.

NEMRC (2022). Seismic Bulletin of the Nepali National Earthquake Monitoring and Research Center [Online]. Available at: http://seismonepal.gov.np.

Replumaz, A., Karason, H., van der Hilst, R.D., Besse, J., Tapponnier, P. (2004). 4-D evolution of SE Asia's mantle from geological reconstructions and seismic tomography. *Earth Planet. Sci. Lett.*, 221(1–4), 103–115.

Rippe, D. and Unsworth, M. (2010). Quantifying crustal flow in Tibet with magnetotelluric data. *Phys. Earth Planet. Int.*, 179, 107–121.

Schulte-Pelkum, V., Monsalve, G., Sheehan, A.F., Shearer, P., Wu, F., Rajaure, S. (2019). Mantle earthquakes in the Himalayan collision zone. *Geology*, 47(9), 815–819.

Styron, R., Taylor, M., Okoronkwo, K. (2010). Database of active structures from the Indo-Asian Collision. *Eos Trans. AGU*, 91(20), 181–182.

Subedi, S., Hetényi, G., Vergne, J., Bollinger, L., Lyon-Caen, H., Farra, V., Adhikari, L.B., Gupta, R. (2018). Imaging the Moho and the Main Himalayan Thrust in Western Nepal with receiver functions. *Geophys. Res. Lett.*, 45, 13222–13230.

Tilmann, F., Ni, J., INDEPTH III Seismic Team (2003). Seismic imaging of the downwelling Indian lithosphere beneath central Tibet. *Science*, 300(5624), 1424–1427.

USGS (2022). ANSS Comprehensive Earthquake Catalog of the United States Geological Survey [Online]. Available at: https://earthquake.usgs.gov/earthquakes/search/.

van der Voo, R., Spakman, W., Bijwaard, H. (1999). Tethyan subducted slabs under India. *Earth Planet. Sci. Lett.*, 171(1), 7–20.

Wei, W., Jin, S., Ye, G., Deng, M., Jing, J., Unsworth, M., Jones, A.G. (2010). Conductivity structure and rheological property of lithosphere in Southern Tibet inferred from super-broadband magnetotelluric sounding. *Sci. China Earth Sci.*, 53(2), 189–202.

Xie, C., Jin, S., Wei, W., Ye, G., Jing, J., Zhang, L., Dong, H., Yin, Y., Wang, G., Xia, R. (2016). Crustal electrical structures and deep processes of the eastern Lhasa terrane in the south Tibetan plateau as revealed by magnetotelluric data. *Tectonophysics*, 675, 168–180.

# 5

# Gravity Observations and Models Along the Himalayan Arc

**Rodolphe Cattin[1], György Hetényi[2], Théo Berthet[3]
and Jamyang Chophel[4]**

[1] *University of Montpellier, France*
[2] *Institute of Earth Sciences, University of Lausanne, Switzerland*
[3] *Department of Earth Sciences, Uppsala University, Sweden*
[4] *Earthquake and Geophysics Division, Department of Geology and Mines,
Thimphu, Bhutan*

## 5.1. Introduction

In 1823, George Everest was appointed as superintendent of the Indian Geodetic Survey. During his field campaigns in northern India, he noted that his gravity measurements led to a surprising conclusion: the Himalayan range is associated with a gravity anomaly that could only be explained by a mass deficiency. This result suggests that this mountain had a much lower density than the surface rocks. Far from being erroneous, his measurements confirmed observations made a century earlier in the Andes: *"Pour revenir aux observations faites sur le Chimborazo, il paraît assez qu'on peut dire en se refermant sur le fait simple, que les montagnes agissent en distance, mais*

*Himalaya, Dynamics of a Giant 1,*
coordinated by Rodolphe Cattin and Jean-Luc Epard.
© ISTE Ltd 2023.

*que leur action est bien moins considérable que le promet la grandeur de leur volume*"[1] (Bouguer 1749).

Since these first discoveries, the amount and quality of gravity measurements along the Himalayan range have continuously increased. Hundreds of gravity field measurements on land (see compilation presented by Hetényi et al. 2016) as well as thousands of space gravity gradiometry[2] data are now available. Together, these observations improve the density distribution assessment under the Himalayan range, and thus better define the three-dimensional geometry of Himalayan and Tibetan lithospheric structures.

In this chapter, after a brief presentation of gravity methods and corrections, we go through the main isostasy models. Next, we show how land gravity measurements are used to explain the mechanical support of the Himalayan topography, characterized by a spatial extension of $\sim$2,500 km in length and peaks exceeding 8,000 m in altitude. Combined with the seismological data described in Volume 1 – Chapter 4, we use gravity measurements to address the north–south flexural rigidity variations of the Indian plate as it deepens under the Tibetan plateau. Finally, we present the space gravity data acquired over the last two decades to extend the study areas associated with land measurements and to increase the depth of investigation.

## 5.2. Methods

### 5.2.1. *Measurements*

It is well known that the period of swing $T$ of a simple pendulum is

$$T = 2\pi\sqrt{\frac{L}{g}} \qquad\qquad [5.1]$$

where $L$ is the length of the pendulum, and $g$ is the local acceleration of gravity[3]. Measurements of the Earth's gravitational acceleration can thus be

---

1. "*To return to the observations made on the Chimborazo, it seems enough to say by closing on the simple fact, that the mountains act in distance, but that their action is much less considerable than the size of their volume promises.*"

2. Gravity gradiometry measurements: measurements of the gravity gradient, i.e. the vertical and horizontal variations of the gravity field measured at two points at a known distance.

3. $g$ is approximately 9.81 m.s$^{-2}$ in Paris.

obtained from the period of a pendulum. During the 17th and 19th centuries, more and more sophisticated pendulums were developed and then became the standard instruments of gravimetry. These pioneering gravity measurements allowed significant advances in understanding the shape of the Earth. They also provided the first information on its composition. The 20th century has seen considerable technological advances. Absolute pendulum gravimeters were replaced by free fall gravimeters, in which a tiny weight is let to fall in a vacuum tank and an optical interferometer measures its acceleration. Today, absolute gravity measurements are obtained with a remarkable accuracy of approximately $\mu$Gal (1 Gal = $10^{-2}$ m.s$^{-2}$), which is equivalent to knowing the value of the gravity field with an accuracy of one part per billion!

Absolute measurements are, however, time-consuming and challenging to implement. Therefore, we favor relative gravity measurements carried out in the field by gravimeters that are easier to transport and to handle. The accuracy of relative measurements is approximately 100 $\mu$Gal, a value that is large enough to detect variations in density associated with geological structures. Most of the data presented in section 5.4 are relative gravity observations acquired in India, along the Himalayan belt, and on the Tibetan plateau. Only a few absolute measurements have been performed to adjust these relative measurements to the previously acquired set of measurements.

Today, the most recent instruments also enable measurement of the gradient of the gravity field from airborne or satellite gradiometers. These measurements are expressed in eotvos, in honor of Baron Loránd Eötvös (1848–1919), pioneer of gravity gradiometry measurements. One eotvos represents a variation of one mGal over a distance of 10 km. In section 5.5, we will see how these new data sets contribute to the study of Himalayan dynamics.

## 5.2.2. Corrections

Before being used, gravity measurements must be corrected. The first correction is related to instrumental drift. Like many measuring instruments, relative gravimeters drift with time, like a watch that is running a little late or early. Hence, two measurements made at the same location at two different times will not be identical. This drift is estimated from repeated gravity measurements. Assuming that this drift is linear with time and that the time elapsed between each measurement is known, it is possible to correct all measurements carried out in a field campaign. The amplitude of this correction depends on the instrument and can reach up to 0.1 mGal per hour.

Gravity measurements are also sensitive to temporal or spatial variations due to external forcings[4]. Thus, the effect of atmospheric pressure changes or tidal deformation of both solid-Earth and ocean-loading must also be corrected (e.g. Longman 1959; Merrian 1992). Although the amplitude of these corrections is quite small ($\sim$ 0.1 mGal), it is essential to apply them to obtain accurate corrected measurements.

The last corrections are due to the shape of the Earth and its topography. Indeed, any gravity measurement depends on its distance from the center of mass of the Earth (Figure 5.1). Due to its rotation, the Earth is not spherical but ellipsoidal with a polar radius of $\sim$ 6,357 km and an equatorial radius of $\sim$ 6,378 km. Even with a homogeneous density distribution in its interior, the gravity at the Earth's surface would be greater at the poles than at the equator. Therefore, a latitudinal correction is required to correct the difference in gravitational acceleration as a function of distance from the equator.

Moreover, the distance from the center of mass of the Earth depends on the topography. Indeed, at the same latitude, a station located at the top of a mountain is farther from the center of mass of the Earth than one located at sea level. Therefore, a correction, called "free-air correction", must be performed proportionally to the measurement elevation. Finally, a mountain is a massive body. This mass also affects the gravity field. A correction, called "Bouguer correction", is made to take into account this effect. The amplitude of these corrections depends strongly on the measurement elevation. In the case of the Himalaya, where some peaks exceed 8,000 m, these corrections can reach up to several hundred mGal. Therefore, their amplitude can be of the same order of magnitude or even greater than the corrected measurements. It is therefore essential to make these corrections very carefully.

To summarize, gravity measurements, corrected for instrumental drift and external forcings, must also be corrected for effects associated with latitude and altitude (topography and relief) before any interpretation in terms of density variations at depth can be made (see Dubois et al. 2022, for more details).

---

4. The term "external forcings" is used to define processes that are not related to the internal dynamics of the Earth

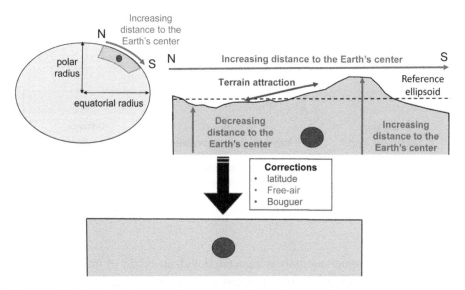

**Figure 5.1.** *Gravimetric corrections associated with a north–south measurement profile in the northern Hemisphere. The latitudinal correction is related to the ellipsoidal shape of the Earth, which induces an increase in the distance from the center of the Earth for the southernmost measurements. The free-air correction is associated with topographic changes that lead to variations in distance from the Earth's center of mass. The topography also induces mass variations and associated gravity changes, which are taken into account via the Bouguer correction. Applying all these corrections allows studying the remaining gravity anomalies associated only with density variations at depth. The density is indicated by a color code, which highlights the existence of a density anomaly at depth. For a color version of this figure, see www.iste.co.uk/cattin/himalaya1.zip*

### 5.2.3. *Anomalies*

The corrected observations (measured field minus corrections) are only sensitive to the attraction of massive bodies associated with density anomalies below the sea level. They highlight bodies having a density contrast compared to a reference Earth model. "Free-air" gravity anomaly is associated with measurements for which latitudinal and free-air corrections are applied. "Bouguer anomaly" is the free-air anomaly with the extra Bouguer correction. These gravity anomalies can be used over several spatial scales, whether to study a cavity of a few meters, a kilometer-long geological structure or a region such as the Himalaya.

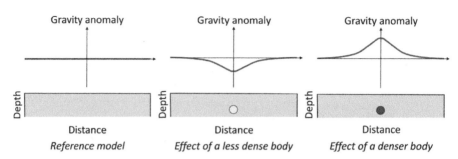

**Figure 5.2.** *Simplified diagram of a Bouguer anomaly profile associated with a mass deficiency or excess related to a buried sphere with low or high density, respectively. The color code gives the density distribution: light (dark) color = low (high) density. For a color version of this figure, see www.iste.co.uk/cattin/himalaya1.zip*

A positive Bouguer anomaly is associated with an excess of mass compared to a reference model. In contrast, a negative anomaly is related to a deficiency of mass below the gravity measurement stations (Figure 5.2). As observed by George Everest, in the Himalaya and Tibet, the negative Bouguer anomaly suggests a mass deficiency below these two regions. The models, allowing to explain this counter-intuitive observation, are presented in the following section.

## 5.3. Isostasy

Maintaining the high reliefs requires an in-depth understanding of the links between topography and mass distribution. To appreciate this problem, let us return to the work of the French astronomer Pierre Bouguer, who, during an expedition to Peru in 1736–1743, showed that the gravity in the Andes is systematically weaker than at sea level (Bouguer 1749). This study and Everest's observations suggest that the high reliefs are associated with a mass deficiency at depth, compared to the surrounding plains.

### 5.3.1. *Local compensation*

In the second half of the 19th century, two theories of the compensation of the reliefs were proposed to explain the observations made by Everest, then by Pratt (1855) in the Himalaya:

– On the one hand, according to John Henry Pratt, the crust under the high reliefs is warmer and therefore less dense than in the adjacent areas (Pratt 1855, 1859). The mass of the topography would thus be locally compensated at depth by a lateral variation of density in the Earth's crust below the topography, without any change in depth of the bottom of the crust (Figure 5.3a).

– For George Biddell Airy, on the other hand, the crust's density does not vary (Airy 1855). Knowing that the crust has a lower density than the mantle, the compensation of the reliefs is then ensured locally by an increase of the crustal thickness (Figure 5.3b).

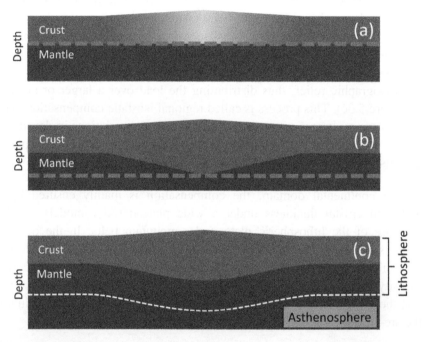

**Figure 5.3.** *Isostasy compensation models. (a) Local compensation in the Pratt hypothesis. The dashed blue line corresponds to the compensation depth, i.e. the depth below which all pressures are hydrostatic. (b) Local compensation in the Airy hypothesis. The dotted blue line corresponds to the compensation depth. (c) Regional compensation provided by the flexure of the lithosphere. The color code gives the density distribution: light (dark) color = low (high) density. For a color version of this figure, see www.iste.co.uk/cattin/himalaya1.zip*

According to these two assumptions, the pressure exerted below the compensation depth[5] is hydrostatic. The weight of each column of the material above the compensation level is constant. The isostatic compensation is then accommodated locally.

### 5.3.2. *Regional compensation*

Based on the plate tectonics theory, we now know that the crust and the top-most part of the upper mantle form a relatively rigid layer called the lithosphere. This layer floats on a weak layer that behaves like a viscous fluid, the asthenosphere. In Finland, following geodetic surveys showing a systematic uplift of his country, Heiskanen proposed a modification of the Airy model accounting for lithospheric rigidity maintaining the reliefs. Indeed, depending on its rigidity, the lithosphere bends more or less under the weight of the topographic relief, thus distributing the load over a larger or smaller area (Figure 5.3c). This process is called regional isostatic compensation. The landforms are no longer compensated only locally by variations in density or thickness of the crust, but regionally by the rigidity of the lithospheric plates (see Watts (2001), for more details).

In the continental domain, the compensation is mainly ensured by a variation in crustal thickness under a wide plateau (Airy model) and by the flexure of the lithospheric plate under mountain belts. In the oceanic domain, isostasy can also be accommodated locally or regionally. The oceanic lithosphere is created from hot mantle rock at a ridge. As the lithosphere cools, its density increases; as a result, it subsides (Pratt's model). In subduction zones, the flexure of the lithosphere plays a significant role in the evolution of the arc–deep-sea trench systems.

Whether the compensation is local or regional, the excess mass at the surface due to the topography is accompanied by a mass deficiency at depth, in agreement with the observations of Pierre Bouguer and George Everest.

---

5. Compensation depth: depth of the crust-mantle boundary in the Pratt model, depth of the thickest crust in the Airy model.

### 5.3.3. *Effective elastic thickness*

The study of the bending of the lithosphere enables the assessment of its elastic properties. Lithospheric rigidity $D$ can be inferred from topographic data and Bouguer anomaly measurements (Figure 5.3b,c). In the case of a non-compensated mountain range, the Bouguer anomaly is zero since no crustal root maintains the relief of mountains. The lithosphere can then be viewed as a highly rigid plate. Conversely, in an area in local isostatic equilibrium, the Bouguer anomaly agrees with gravity associated with Airy's model. In that case, the lithosphere will be regarded as having zero rigidity.

RHEOLOGY OF THE LITHOSPHERE.    Rheology is the study of the deformation and flow of matter under the effect of thermo-mechanical stresses. Rheological profiles describe the mechanical behavior of the lithosphere with depth. These profiles involve specific laws (Bayerlee's law, creep law), which define the strength of the lithosphere as a function of depth, and its brittle or ductile behavior. These profiles are obtained from experimental measurements in the laboratory and field observations, including petrological and structural studies, location of seismicity at depth, and joint analysis of gravity anomalies and topography.

The flexural rigidity of the lithosphere varies between $10^{20}$ and $10^{25}$ N.m (e.g. Watts 2001). These values are difficult to understand. We, therefore, favor the use of the effective elastic thickness, which corresponds to the thickness of a thin plate with a rigidity $D$. This thickness $h$ is defined by

$$h = \left( \frac{12D(1 - \nu^2)}{E} \right)^{1/3} , \qquad\qquad [5.2]$$

where $E$ and $\nu$ are the Young modulus and the Poisson ratio of the plate material, respectively. In the oceanic domain, the effective elastic thickness is directly correlated to the thermal age of the lithosphere. The older the lithosphere is, the colder it is and the greater its rigidity and equivalent thickness. This thickness then corresponds to the depth of the 450–600°C isotherm, which is used to define the base of the oceanic lithosphere. In the continental domain, the physical meaning of this thickness is more complex. The continental lithosphere consists of an upper and lower crust and part of the upper mantle. The associated rigidity is then calculated for a composite plate. The effective elastic thickness $h_{eff.}$ of the lithosphere is obtained from the elastic thickness of each layer:

$$h_{eff.} = \left(h_{upper\ crust}^3 + h_{lower\ crust}^3 + h_{mantle}^3\right)^{1/3}. \qquad [5.3]$$

Numerical approaches are now available to calculate the deformations associated with topographic loading by considering the distribution of elastic parameters $E$ and $\nu$ and the density $\rho$ in the lithosphere. By testing many models with different parameter values, it is, therefore, possible to use $h_{eff.}$ as a parameter to improve the assessment of lithospheric rheology.

## 5.4. Flexure of the Indian plate

Over the last four decades, numerous research studies have been carried out on the flexure of the Indian plate in the context of the India–Asia collision (e.g. Lyon-Caen and Molnar 1983; Burov and Watts 2006; Hetényi et al. 2006; Berthet et al. 2013).

Initially focused on central Nepal, gravity studies showed that the Himalaya was not locally compensated. Part of the load of the highest reliefs had to be maintained by the rigidity of the Indian lithosphere. Since then, several gravimetric measurement campaigns have been carried out to verify this result and test its validity along the entire Himalayan chain.

### 5.4.1. *Gravity anomaly across the Himalayan belt*

A unique set of land gravity data is now available (Figure 5.4). It consists of about 20 absolute measurements and more than 2,700 relative measurements made in India, Bangladesh, Nepal, Bhutan and Tibet (see the compilation of Hetényi et al. (2016)). The average accuracy of relative measurements is less than 10 mGal. These observations cover more than 2,000 km along the Himalayan arc. They show a strong Bouguer anomaly decrease (>500 mGal) between India and Tibet. This decrease is localized over a zone, which is 600 km wide, including the northernmost part of the Indian peninsula, the whole Himalayan range and the southern Tibetan plateau.

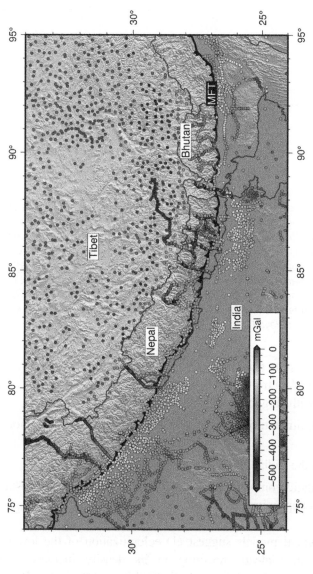

**Figure 5.4.** *Bouguer anomaly map of the Himalaya and surrounding regions. (e.g. Das et al. 1979; Sun 1989; Banerjee 1998; Martelet et al. 2001; Cattin et al. 2001; Tiwari et al. 2006; Hammer et al. 2013; Berthet et al. 2013, Poretti (personnal communication) and the available data from the International Gravimetric Bureau). Color circles give the location of gravity measurements and the associated Bouguer anomaly. Boundaries of countries, geographic regions and the Main Frontal Thrust (MFT) are shown as a reference. For a color version of this figure, see www.iste.co.uk/cattin/himalaya1.zip*

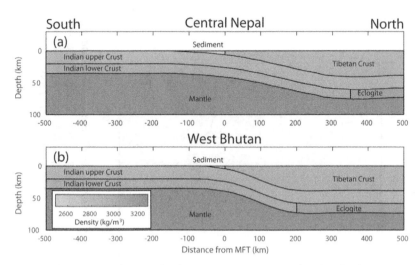

**Figure 5.5.** *Geometry of lithospheric structures and density distribution across the Himalayan range. These models are obtained from Bouguer anomaly data (Figures 5.4 and 5.6), seismological observations and borehole measurements across the Himalaya, from India to Tibet. They account for the sedimentary basin associated with the foreland basin, the Tibetan crust and the Indian plate, which is composed of three layers: the upper crust, the lower crust (eclogitized underneath Tibet) and the lithospheric mantle. Distances are given from the Himalayan Frontal Thrust (MFT). (a) Model obtained for central Nepal (Berthet et al. 2013). (b) Model obtained for western Bhutan (Hammer et al. 2013). For a color version of this figure, see www.iste.co.uk/cattin/himalaya1.zip*

Combined with results from data of seismological networks deployed over the entire region and data from deep boreholes in India, these gravity variations can be associated with (Figure 5.5):

– the foreland sedimentary basin, whose depth can locally reach more than 8 km;

– the flexure of the Indian lithosphere under the Himalaya;

– the local isostatic compensation of the Tibetan plateau by a crustal root with a thickness of approximately 35 km.

Thermo-mechanical models suggest (1) eclogitization of the lower crust beneath the Tibetan Plateau, resulting in its density increase and (2) north–south variations in the effective elastic thickness of the Indian Plate (e.g. Cattin et al. 2001; Hetényi et al. 2006, 2007). The Indian plate is relatively cold before it sinks under the mountain belt. The upper crust, lower crust and mantle are then strongly coupled, leading to an effective elastic thickness of approximately 70 km. Further north, the deepening of the Indian plate, and the associated heating, lead to a substantial decrease in the crustal rigidity. The effective elastic thickness of the Indian lithosphere is reduced by half to approximately 30 km. The topographic load is then maintained by the rigidity of the Indian lithospheric mantle alone.

### 5.4.2. *Along-strike variation between Nepal and Bhutan*

The tectonic units are remarkably continuous along the 2,500 km long arc of the Himalaya. The observed gravity anomalies can therefore be interpreted using two-dimensional models perpendicular to the structures of the Himalayan arc (Berthet et al. 2013). However, a detailed analysis reveals the existence of lateral variations along the Himalaya (Hammer et al. 2013). We focus on the region between central Nepal and eastern Bhutan, where the Bouguer anomaly is well-documented across the mountain belt (Figure 5.4). There we can see the following longitudinal variations in gravity profiles (Figure 5.6):

– South of the frontal thrust, in the foreland basin, the Bouguer anomaly is approximately -300 mGal in Nepal, compared to approximately -100 mGal in Bhutan. This difference suggests a larger and deeper sedimentary basin in southern Nepal than in south of Bhutan (Figure 5.5).

– About 100–200 km north of the frontal thrust, the anomaly is more pronounced in Bhutan than Nepal, suggesting a more rapid deepening of the Indian plate beneath Bhutan compared to Nepal (Figure 5.5).

These along-strike variations can be related to an eastward decrease in the effective elastic thickness from approximately 30 km in Nepal to approximately 20 km in Bhutan. This finding thus suggests strong rheological variations of the Indian plate between Nepal and Bhutan. It is in agreement (1) with seismological studies (see Volume 1 – Chapter 4) have detected an active fault cutting the Himalaya and the foreland basin between these two areas, the Dhubri–Chungthang fault (Diehl et al. 2017) and (2) with geodetic studies that propose relative motion between the India plate and the block associated with the Shillong Plateau, located east of this fault (e.g. Vernant et al. 2014).

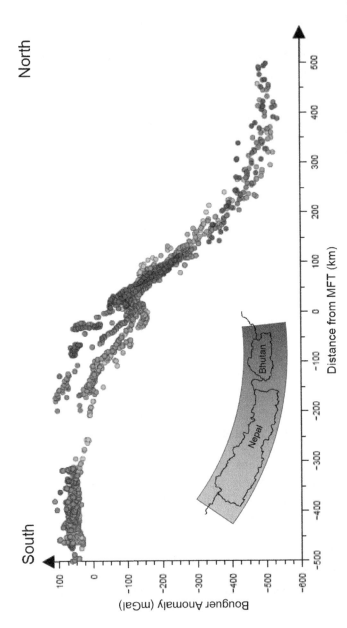

**Figure 5.6.** *Bouguer anomaly profiles between western Nepal and eastern Bhutan across the Himalayan range, perpendicular to the Main Frontal Thrust (MFT). The color scale indicates the location of the gravity measurements along the Himalayan arc: Yellow, West Nepal - Red, East Bhutan. For a color version of this figure, see www.iste.co.uk/cattin/himalaya1.zip*

## 5.5. Satellite data contribution

Land gravity measurements provide valuable information on the geometry and rheology of lithospheric structures. However, field conditions associated with high altitude, limited road network, transboundary issues and extreme climatic events make the acquisition of these measurements often difficult, sometimes impossible. Thus, many areas such as western Nepal, Himachal Pradesh in the west or Arunachal Pradesh in the east remain poorly or not surveyed (Figure 5.4). Consequently, the coverage of land measurements remains insufficient to study lateral variations in detail, over the entire Himalayan region.

The first solution is to set up airborne acquisition campaigns. This type of measurement exists for part of Nepal, but these government data are not available to the scientific community. Furthermore, it seems illusory to obtain flight authorizations over territories disputed by India and China. Even then, it would be impossible to get complete coverage because of the steep terrain and flight conditions.

Therefore, the only possibility to complete the land data is to go even higher, using satellite measurements.

### 5.5.1. *Gravity measurements from space*

The first space measurements of the Earth's gravity field were made in the late 1950s by studying the orbits of artificial satellites. One of the first dedicated campaigns is the GRACE[6] mission launched in 2002. The GRACE mission provided global information on the gravity field and its temporal variations. The method is based on differential acceleration measurements between two satellites flying at an altitude of approximately 400 km. This altitude gives a 600–1,000 km resolution at the Earth's surface, which is still too high to study areas such as the Himalayan arc.

In 2009, the launch of the GOCE satellite[7] flying at an altitude between 225 km and 265 km has overcome this limitation by yielding a resolution below 100 km on the ground.

---

6. GRACE: Gravity Recovery And Climate Experiment.
7. GOCE: Gravity field and steady-state Ocean Circulation Explorer.

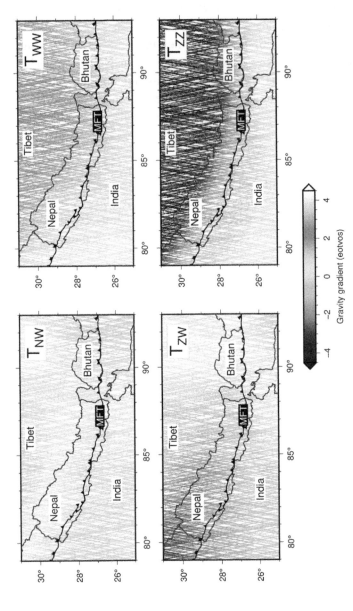

**Figure 5.7.** *Map of gravity gradients including topographic corrections from the spatial gravity mission GOCE. Color dots represent the data along the satellite orbits at an altitude between approximately 225 km and 265 km. $T_{ij}$ is the $ij$ component of gravity gradient tensor. $T_{NW}$, $T_{WW}$ and $T_{ZW}$ are associated with the partial derivative of the three gravity components in the west direction. $T_{ZZ}$ is the partial derivative of $g_z$ in the vertical direction. Borders of countries and the main tectonic structures are shown as a reference. For a color version of this figure, see www.iste.co.uk/cattin/himalaya1.zip*

On board of the GOCE satellite, three perpendicular gradiometers measure the complete gradient tensor $T$, which is symmetric and its trace is zero:

$$T = \begin{pmatrix} T_{WW} & T_{WN} & T_{WZ} \\ T_{NW} & T_{NN} & T_{NZ} \\ T_{ZW} & T_{ZN} & T_{ZZ} \end{pmatrix} \qquad [5.4]$$

where $T_{ij} = \frac{\partial g_i}{\partial x_j}$, with $g_i$ being the component of the gravity field in the $i$ direction ($W$: west; $N$: north; $Z$: vertical) and $x_j$ being the distance in the $j$ direction. $T_{NW}$, $T_{WW}$ and $T_{ZW}$ are the partial derivatives of $\vec{g}(g_N, g_W, g_Z)$ in the west direction. $T_{ZZ}$ is the partial derivative of $g_Z$ in the vertical direction. As previously mentioned, $T$ is expressed in eotvos[8].

Land gravity measurements and space gravity gradiometry are complementary data. On the one hand, GOCE measurements are sensitive to the three components of the gravity field, while on the ground, only the vertical component is measured. Gravity gradiometry observations better characterize the geometry of anomalies and associated bodies compared to land gravity measurements. The more homogeneous coverage of satellite data also allows a better characterization of the long wavelengths of the gravity field. On the other hand, land measurements are closer to the source of anomalies. They are more sensitive to slight density variations and more relevant to study localized gravity variations.

To maximize the signal-to-noise ratio, we consider the period between August 2012 and September 2013, for which the satellite operated in low orbit at an altitude as low as 224 km. The GOCE dataset consists of 17,500 measurements for the five independent $T$ components from western Nepal to eastern Bhutan. The estimated error is $\sim 0.1$ E. These measurements show for $T_{NW}$ small variations ($< 0.5$ E) and for $T_{ZW}$ significant lateral variations (Figure 5.7). In the Tibetan plateau, while positive values ($>2.5$ E) are observed for $T_{WW}$, $T_{ZZ}$ exhibits very negative values ($<-3.5$ E).

### 5.5.2. *Towards a three-dimensional image*

Combined with seismological (e.g. Schulte-Pelkum et al. 2005; Nábělek et al. 2009; Singer et al. 2017) and land gravity measurements, GOCE data

---

8. 1 eotvos = $10^{-9}$ s$^{-2}$. Its symbol is E.

provide an unprecedented level of detail for imaging lithospheric structures across the Himalayan range.

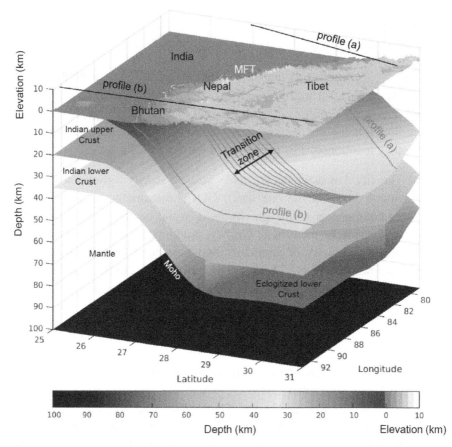

**Figure 5.8.** *Three-dimensional model showing the geometry of the Indian lithosphere under the Himalayan range and the Tibetan plateau. It is obtained from all available data, including land and space-based gravity observations. Profiles (a) and (b) correspond to two-dimensional models obtained in central Nepal and west Bhutan, respectively (Figure 5.5). The transition zone located at the western boundary of Bhutan is related to an abrupt change in the geometry of the Indian plate. The density distribution is similar to the one used in the two-dimensional models. For a color version of this figure, see www.iste.co.uk/cattin/himalaya1.zip*

We have shown in section 5.4.2 that the lateral variations observed between Nepal and Bhutan could be associated with a difference in the mechanical behavior of the Indian plate. Here, we use all data sets jointly to better constrain this difference, i.e. where is the transition zone between the two-dimensional geometries shown in Figure 5.5? What is its width? (Cattin et al. 2021).

The inversion of all land and satellite gravity data provides a three-dimensional image of the lithospheric structures of the region (Figure 5.8). The results suggest a ca. 10 km wide transition, relatively narrow, between Sikkim and Bhutan. This abrupt segmentation is supported by structural observations and could be related to seismically active structures such as the Dhubri–Chungthang fault that cuts the Indian plate beneath the Himalaya. It could also be associated with currently inactive tectonic structures such as the Madhupur fault in the foreland and the Yadong–Gulu rift in southern Tibet.

The obtained transition zone is narrow enough to possibly prevent seismic rupture propagation across this boundary between Nepal and Bhutan. This abrupt geometry could result in the seismic segmentation of the Main Himalayan Thrust and potentially put a higher limit to the size of large earthquakes along the Himalayan arc. These hypotheses now need to be studied in further detail. They could be critical components of future seismic hazard models in this tectonically highly active area (see Volume 3 – Chapters 5 and 6).

## 5.6. Conclusion

Since the 19th century, the Himalaya have been an area of pioneering scientific work in studying the Earth's gravity field. The isostasy models of John Henry Pratt and George Biddell Airy, developed to understand the compensation of Himalayan highlands, have been successfully applied to many regions worldwide. Similarly, the concepts of flexural rigidity and effective elastic thickness, used to describe the deformation of the Indian plate, are now commonly used to better characterize the rheology of the lithosphere in both continental and oceanic domains.

Gravity and gradiometry measurements, on the land and from space, are complementary to seismological observations. Together, these data allow us to better image the lithospheric structures associated with the India–Asia collision. In particular, they suggest a complex and three-dimensional geometry of the Himalayan arc with steep segmentations, such as the one located at the western border of Bhutan. These new images are crucial for estimating the seismic hazard of this very active region (e.g. Le Roux-Mallouf et al. 2020; Stevens et al. 2020, and Volume 3 – Chapter 6). Indeed, these transition zones could control the seismic segmentation of the Himalayan range by limiting the size of earthquakes (e.g. Hetényi et al. 2016).

These results are to be confirmed: on the one hand, by increasing the coverage of land measurements, especially in western Nepal, and on the other hand, by improving space observations. Launched in 2018, the GRACE-FO[9] mission aims to characterize the spatio-temporal variations of the gravity field. These data will allow the study of interactions between external and internal forcings in the Himalayan region, where the movement of water masses on the Earth's surface is strongly affected by monsoon cycles. Future gravity field mapping campaigns are underway with innovative absolute gravimeters, which are based on atomic interferometry techniques with cold atoms.

These scientific advances and field investigations will allow us to further improve our knowledge of the Himalayan dynamics, and thus to tackle major societal challenges such as access to water resources and land use planning in this region, where extreme climatic and tectonic events occur.

## 5.7. References

Airy, G.B. (1855). On the computation of the effect of the attraction of mountain-masses, as disturbing the apparent astronomical latitude of stations in geodetic surveys. *Phil. Trans. Roy. Soc. London*, 145, 101–104.

Banerjee, P. (1998). Gravity measurements and terrain corrections using a digital terrain model in the NW Himalaya. *Comp. Geosci.*, 24, 1009–1020.

Berthet, T., Hetényi, G., Cattin, R., Sapkota, S.N., Champollion, C., Kandel, T., Doerflinger, E., Drukpa, D., Lechmann, S., Bonnin, M. (2013). Lateral uniformity of India Plate strength over central and eastern Nepal. *Geophys. J. Int.*, 195(3), 1481–1493.

Bouguer, P. (1749). *La figure de la Terre, déterminée par les observations de messieurs Bouguer et de la Condamine, de l'Académie royale des sciences, envoyés par ordre du roy au Pérou, pour observer aux environs de l'équateur*. Charles-Antoine Jomdert, Paris.

Burov, E.B. and Watts, A.B. (2006). The long-term strength of continental lithosphere: "Jelly sandwich" or "crème brûlée"? *GSA Today*, 16(1), 4–10.

---

9. GRACE-FO: Gravity Recovery and Climate Experiment Follow-On.

Cattin, R., Martelet, G., Henry, P., Avouac, J.P., Diament, M., Shakya, T.R. (2001). Gravity anomalies, crustal structure and thermo-mechanical support of the Himalaya of Central Nepal. *Geophys. J. Int.*, 147(2), 381–392.

Cattin, R., Berthet, T., Hetényi, G., Saraswati, A., Panet, I., Mazzotti, S., Cadio, C., Ferry, M. (2021). Joint inversion of ground gravity data and satellite gravity gradients between Nepal and Bhutan: New insights on structural and seismic segmentation of the Himalayan arc. *Phys. Chem. Earth A/B/C*, 123, 103002.

Das, D., Mehra, G., Rao, K.G.C., Roy, A.L., Narayana, M.S. (1979). Bouguer, free-air and magnetic anomalies over northwestern Himalaya. Himalayan Geology seminar, Section III, Oil and Natural Gas Resources. *Geol. Surv. India Misc. Publ.*, 41, 141–148.

Diehl, T., Singer, J., Hetényi, G., Grujic, D., Giardini, D., Clinton, J., Kissling, E., GANSSER Working Group (2017). Seismotectonics of Bhutan: Evidence for segmentation of the Eastern Himalayas and link to foreland deformation. *Earth Planet. Sci. Lett.*, 471, 54–64.

Dubois, J., Diament, M., Cogné, J.-P., Moquet, A. (2022). *Géophysique*. Dunod, Paris.

Hammer, P., Berthet, T., Hetényi, G., Cattin, R., Drukpa, D., Chophel, J., Lechmann, S., Moigne, N.L., Champollion, C., Doerflinger, E. (2013). Flexure of the India plate underneath the Bhutan Himalaya. *Geophys. Res. Lett.*, 40(16), 4225–4230.

Hetényi, G., Cattin, R., Vergne, J., Nábělek, J. (2006). The effective elastic thickness of the India Plate from receiver function imaging, gravity anomalies and thermomechanical modelling. *Geophys. J. Int.*, 167(3), 1106–1118.

Hetényi, G., Cattin, R., Brunet, G., Vergne, J., Bollinger, L., Nábělek, J., Diament, M. (2007). Density distribution of the India plate beneath the Tibetan Plateau: Geophysical and petrological constraints on the kinetics of lower-crustal eclogitization. *Earth Planet. Sci. Lett.*, 264, 226–244.

Hetényi, G., Cattin, R., Berthet, T., Le Moigne, N., Chophel, J., Lechmann, S., Hammer, P., Drukpa, D., Sapkota, S.N., Gautier, S. et al. (2016). Segmentation of the Himalayas as revealed by arc-parallel gravity anomalies. *Sci. Rep.*, 6(1), 1–10.

Le Roux-Mallouf, R., Ferry, M., Cattin, R., Ritz, J.F., Drukpa, D., Pelgay, P. (2020). A 2600-year-long paleoseismic record for the Himalayan Main Frontal Thrust (western Bhutan). *Solid Earth*, 11(6), 2359–2375.

Longman, I.M. (1959). Formulas for computing the tidal acceleration due to the Moon and the Sun. *J. Geophys. Res.*, 64(12), 2351–2355.

Lyon-Caen, H. and Molnar, P. (1983). Constraints on the structure of the Himalaya from an analysis of gravity anomalies and a flexural model of the lithosphere. *J. Geophys. Res.*, 88, 8171–8191.

Martelet, G., Sailhac, P., Moreau, F., Diament, M. (2001). Characterization of geological boundaries using 1-D wavelet transform on gravity data: Theory and application to the Himalayas. *Geophysics*, 66, 1116–1129.

Merriam, J.B. (1992). Atmospheric pressure and gravity. *Geophysical Journal International*, 109, 231–244.

Nábělek, J., Hetényi, G., Vergne, J., Sapkota, S., Kafle, B., Jiang, M., Su, H., Chen, J., Huang, B.S. (2009). Underplating in the Himalaya-Tibet collision zone revealed by the Hi-CLIMB experiment. *Science*, 325(5946), 1371–1374.

Pratt, J.H. (1855). On the attraction of the Himalaya Mountains and of the elevated regions beyond them, upon the plumb line in India. *Phil. Trans. Roy. Soc. London*, 145, 53–100.

Pratt, J.H. (1859). On the deflection of the plumb-line in India, caused by the attraction of the Himmalaya mountains and of the elevated regions beyond; and its modification by the compensating effect of a deficiency of matter below the mountain mass. *Phil. Trans. Roy. Soc. London*, 149, 745–796.

Schulte-Pelkum, V., Monsalve, G., Sheehan, A., Pandey, M.R., Sapkota, S., Bilham, R., Wu, F. (2005). Imaging the Indian subcontinent beneath the Himalaya. *Nature*, 435(7046), 1222–1225.

Singer, J., Kissling, E., Diehl, T., Hetényi, G. (2017). The underthrusting Indian crust and its role in collision dynamics of the Eastern Himalaya in Bhutan: Insights from receiver function imaging. *J. Geophys. Res. Solid Earth*, 122, 1152–1178. doi:10.1002/2016JB013337.

Stevens, V.L., De Risi, R., Le Roux-Mallouf, R., Drukpa, D., Hetényi, G. (2020). Seismic hazard and risk in Bhutan. *Nat. Hazards*, 104, 2339–2367.

Sun, W. (1989). *Bouguer Gravity Anomaly Map of the People's Republic of China*. Chin. Acad. Geoexploration, Beijing.

Tiwari, V.M., Vyghreswara, R., Mishra, D.C., Singh, B. (2006). Crustal structure across Sikkim, NE Himalaya from new gravity and magnetic data. *Earth Planet. Sci. Lett.*, 247, 61–69.

Vernant, P., Bilham, R., Szeliga, W., Drukpa, D., Kalita, S., Bhattacharyya, A.K., Gaur, V.K., Pelgay, P., Cattin, R., Berthet, T. (2014). Clockwise rotation of the Brahmaputra Valley relative to India: Tectonic convergence in the eastern Himalaya, Naga Hills and Shillong Plateau. *J. Geophys. Res.* doi:10.1002/2014JB011196.

Watts, A.B. (2001). *Isostasy and Flexure of the Lithosphere*. Cambridge University Press.

# 6

# Topographic and Thermochronologic Constraints on the Himalayan Décollement Geometry

Peter A. VAN DER BEEK[1], Rasmus C. THIEDE[2],
Vineet K. GAHALAUT[3] and Taylor F. SCHILDGEN[1,4]

[1] *University of Potsdam, Germany*
[2] *Christian-Albrecht University, Kiel, Germany*
[3] *CSIR–National Geophysical Research Institute, Hyderabad, India*
[4] *GFZ German Research Centre for Geoscience, Potsdam, Germany*

## 6.1. Introduction

The Himalaya is the largest mountain belt on Earth, stretching 2,500 km from the Nanga Parbat syntaxis in the northwest to the Namche Barwa syntaxis in the southeast. Knowledge of the mountain belt has traditionally been built up from expeditions in its different sectors, with Gansser (1964) attempting the first "modern" synthesis and map of the entire mountain range. In this and subsequent syntheses of Himalayan structure (Le Fort 1975; Hodges et al. 2000; Yin and Harrison 2000; Kapp and DeCelles 2019), emphasis has been put on the lateral continuity of major faults and litho-tectonic units across

*Himalaya, Dynamics of a Giant 1*,
coordinated by Rodolphe CATTIN and Jean-Luc EPARD.
© ISTE Ltd 2023.

strike, as discussed in detail in Volume 1 – Chapter 3, giving rise to the notion of a largely cylindrical structure of the mountain belt along its length (Figure 6.1).

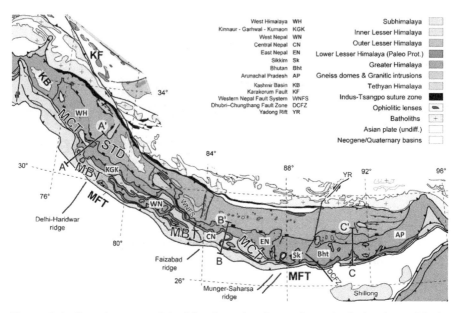

**Figure 6.1.** *Overview map of the Himalaya showing main geological units and faults, representative cross-sections shown in Figure 6.2 and regions discussed in the text. Basement ridges in underthrusting Indian plate are indicated by dark blue lines. Map is extensively modified from Hintersberger et al. (2011), with added data from Grujic et al. (2011), Mandal et al. (2019) and Dal Zilio et al. (2021). For a color version of this figure, see www.iste.co.uk/cattin/himalaya1.zip*

The recognition that the first-order structure of the Himalaya is controlled by underthrusting of the Indian subcontinent below Tibet goes back to Argand (1924) and is central to all current models of Himalayan tectonics. Structural models (e.g. Schelling and Arita 1991; Srivastava and Mitra 1994; DeCelles et al. 2001, 2016; McQuarrie et al. 2008), supported by geophysical data (e.g. Zhao et al. 1993; Schulte-Pelkum et al. 2005; Nábělek et al. 2009; Caldwell et al. 2013), show that the large south-vergent thrusts of the Himalaya, i.e. the Main Central Thrust (MCT), Main Boundary Thrust (MBT) and Main Frontal Thrust (MFT), branch off a single crustal-scale décollement or megathrust: the Main Himalayan Thrust (MHT; Figure 6.2). This feature is the plate interface between India and Asia and the source of major earthquakes in the Himalaya

(Seeber et al. 1981; Bilham et al. 2001; Bilham 2019, and Volume 3 – Chapter 5). The MHT is fundamental to the evolution of the entire Himalaya, as it controls not only its structure and kinematics, but also both the large-scale surface morphology and erosion patterns.

Above the MHT, the Lesser Himalaya often shows an imbricate duplex or antiformal-stack geometry, which uplifts, folds and generally causes regional northward tilting of the overlying Greater and Tethyan Himalaya. Greater Himalayan klippen are preserved in some areas south of the duplexes (e.g. Kathmandu klippe, central Nepal; Figure 6.2). Although the existence of the Lesser Himalayan duplex is generally accepted, there is little agreement on the timing of its formation, the significance of its roof thrust, known as the Ramgarh–Munsiari Thrust (RMT), or the relative timing of deformation along the MCT, RMT and MBT.

Seeber and Gornitz (1983) were the first to link the surface geomorphology (in their case looking at river gradients) to the structure and kinematics of the Himalaya. Detailed studies of Himalayan topography started in the early 2000s, with the advent of large-scale remotely sensed topography datasets. These studies have led to two major insights: (1) there are significant lateral variations in characteristic across-strike topography of the Himalaya, with regions showing a linear increase in topography from the mountain front to the Tibetan Plateau (e.g. Bhutan, western Nepal) alternating with regions that show an abrupt step in topography, associated with the highest relief, the highest slopes and the highest (>8-km) peaks (e.g. central Nepal) (Duncan et al. 2003; Bookhagen and Burbank 2006) and (2) these lateral topographic variations are linked to characteristic precipitation patterns: regions with a linear increase in elevation show a single broad precipitation peak whereas regions with a step change in topography show a double peak in precipitation, with a strong maximum at the topographic step (Bookhagen and Burbank 2006, 2010).

Lateral variations in crustal structure along the belt have also been recognised from compilations of seismic (Priestley et al. 2019, and Volume 1 – Chapter 4), gravity (Godin and Harris 2014, and Volume 1 – Chapter 5) and geological (Gahalaut and Kundu 2012; Hubbard et al. 2021) data. These lateral variations are argued to play a role in delimiting the rupture lengths of great and major Himalayan earthquakes (Gahalaut and Kundu 2012; Godin and

Harris 2014; Dal Zilio et al. 2021; Hubbard et al. 2021, Volume 3 – Chapters 5 and 6). However, the links between these different observations of seismicity, geomorphology and crustal structure, as well as the timescales to which they pertain, have not been fully clarified yet.

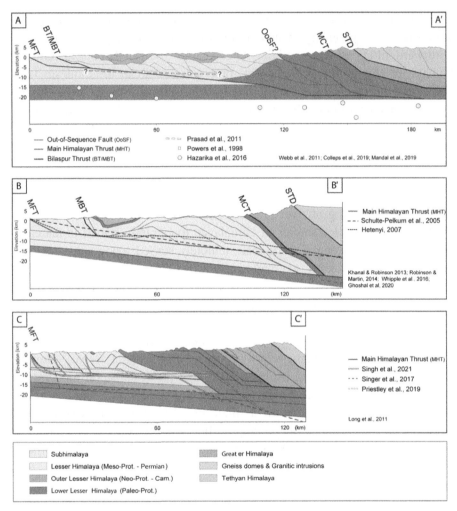

**Figure 6.2.** *Three representative structural cross-sections based on structural data and seismic imaging: (A–A') north-western Himalaya; (B–B') central Nepal Himalaya; (C–C') Bhutan Himalaya. References for the structural cross-sections are shown in the lower-right corner of each panel; constraints on the geometry of the MHT from seismic data are shown in the legend. For a color version of this figure, see www.iste.co.uk/cattin/himalaya1.zip*

Millennial to million-year timescale erosion or exhumation patterns within the mountain belt may provide links between the short- and long-timescale observations outlined above (see Volume 3 – Chapter 2). Thermochronology data have shown that the steep and high-relief regions with strong precipitation peaks along the southern front of the Greater Himalaya are exhuming rapidly, as indicated by very young thermochronologic ages, for instance, in NW India (Thiede et al. 2004, 2005, Sutlej transect) and central Nepal (Wobus et al. 2003; Robert et al. 2009), whereas regions with a more linear increase in topography do not show such young ages, indicating less rapid exhumation, for instance in Bhutan (Grujic et al. 2006), western Nepal (van der Beek et al. 2016) or parts of the north-western Himalaya (Thiede et al. 2017) (Figure 6.1). This spatial correlation between precipitation, morphology and exhumation patterns has led to significant speculation and controversy concerning the potential long-term coupling between tectonics, climate and erosion: some authors suggested that focused precipitation has led to focused exhumation and sustained out-of-sequence thrusting at the steep topographic transition zones (Wobus et al. 2003, 2005, 2006a; Hodges et al. 2004), while others have argued that the precipitation patterns represent a passive response to topographic growth, which together with rapid exhumation is driven by tectonics due to the geometry and kinematics of the MHT (Burbank et al. 2003; Bollinger et al. 2006; Robert et al. 2009; Herman et al. 2010; Godard et al. 2014).

These observations and correlations have led to a number of important questions:

(1) What is the nature of the coupling between tectonics, erosion and climate in the Himalaya; do we only record a one-way coupling or do we see feedbacks?

(2) What controls rapid exhumation in the Himalaya? Are rapidly eroding zones linked to rock uplift above crustal-scale ramps in the MHT, active duplexing, or is there evidence for out-of-sequence faulting?

(3) How can we document and model the lateral variations and segmentation in Himalayan structure, and what are the potential links to seismogenic segmentation (Hubbard et al. 2016; Dal Zilio et al. 2021)?

In this chapter, we address these questions, using the observed topography and morphology of the mountain belt as well as inferred spatio-temporal variations in long-term erosion/exhumation rates to infer the kinematics of mountain building and the role of (lateral variations in) the MHT. We also discuss the complementarity of this approach with geophysical datasets. We first provide a brief background to the methods used: quantitative geomorphic analysis, cosmogenic nuclides, thermochronology and thermo-kinematic modeling. We then review a number of case studies addressing different regions along the Himalayan arc: the western and central Nepal Himalaya, the latter of which has long been treated as the "type section" for Himalayan tectonics; the western Himalaya of northwest India; and the eastern Himalaya of Bhutan and Arunachal Pradesh. Finally, we attempt to bring these different observations together, discuss evidence for or against out-of-sequence thrusting in the different regions, and interpret the observed lateral variations in terms of controls on the structure and kinematics of the mountain belt.

## 6.2. Methods

### 6.2.1. *Quantitative geomorphic analysis*

The relief of mountain belts, in particular the fluvial relief, contains information on tectonic uplift rates modulated by climatic and lithologic imprints (e.g. Lavé and Avouac 2001; Wobus et al. 2006b; Kirby and Whipple 2012). It has been recognized for several decades that rivers flowing across the Himalaya often display very steep middle reaches, approximately where they cross the MCT zone between the Greater and Lesser Himalaya; this observation was interpreted as reflecting rapid uplift in that area (Seeber and Gornitz 1983; Brookfield 1998).

The basis of most modern analyses of tectonics from fluvial relief is the stream-power erosion law, which states that fluvial incision/erosion rates ($\dot{e}$) depend on drainage area ($A$; a proxy for water discharge) and channel slope $S$ to the powers $m$ and $n$, respectively:

$$\dot{e} = K \times A^m \times S^n, \tag{6.1}$$

in which $K$ is a parameter that encompasses climatic, lithologic and hydraulic-geometry influences (Whipple 2004; Lague 2014).

By assuming topographic steady state, meaning that (1) the river profile does not change with time and (2) erosion rate equals uplift rate ($\dot{e} = \dot{U}$), the stream-power equation can be rewritten as a power-law relation between slope and drainage area:

$$S = k_s \times A^{-\theta}, \tag{6.2}$$

where

$$k_s = \left(\frac{\dot{U}}{K}\right)^{-\frac{1}{n}}$$
[6.3]

is known as the "steepness index" and $\theta = m/n$ is the concavity. The steepness index thus expresses the channel slope normalized by drainage area and contains information on rock-uplift rate, modulated by the controls on $K$. This analysis has been widely used to map out regional variations in rock-uplift rate; a well-known example from the central Nepal Himalaya (Kirby and Whipple 2012) is shown in Figure 6.3. This example illustrates the characteristic high steepness-index values across the MCT zone, suggesting enhanced rock-uplift rates in that area.

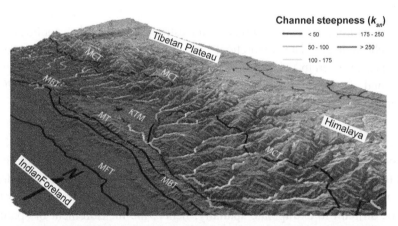

**Figure 6.3.** *Normalized channel-steepness index for main rivers and their tributaries in the central Nepal Himalaya (modified from Kirby and Whipple 2012). Note the strong increase in steepness index at the physiographic transition between the Lesser and Greater Himalaya, corresponding in part to the MCT zone. Major thrusts are also indicated; abbreviations as in Figure 6.1 except MT: Mahabarat Thrust; KTM: Kathmandu. For a color version of this figure, see www.iste.co.uk/cattin/himalaya1.zip*

Although the stream-power approach is attractive because of its simplicity and intuitive link between fluvial relief and tectonic uplift, there are potential issues linked to its use in the Himalayan context. These include: (1) the approach assumes a simple power-law relationship between discharge and drainage area, which is not necessarily true in regions of strong spatial (orography) and temporal (monsoonal rainfall) variations in precipitation like the Himalaya (Bookhagen and Burbank 2010; Andermann et al. 2012); (2) the stream-power law does not take into account channel-width variations that can be important (Lavé and Avouac 2001); and (3) it does not incorporate channel-incision thresholds or the role of sediments (Lavé and Avouac 2001; Lague 2014). More elaborate analyses that take some or all of these effects into account require additional information to calculate specific stream power (taking into account discharge and width variations; Scherler et al. 2014; van der Beek et al. 2016) or excess stream power (also taking into account thresholds; Lavé and Avouac 2001).

## 6.2.2. *Measures of erosion at different timescales: cosmogenic nuclides and thermochronology*

Measuring erosion rates at longer timescales has been mostly based on two isotope-based methods: terrestrial cosmogenic nuclides and thermochronology. The theoretical and methodological bases for these methods are treated in Volume 3 – Chapter 2 and can also be found in numerous books and synthesis papers (e.g. Reiners and Brandon 2006; Dunai 2010; Reiners et al. 2017).

### 6.2.2.1. *Cosmogenic nuclides*

Cosmogenic nuclides provide information on erosion rates on relatively short (typically millennial) timescales and small (<1 m) spatial scales. Bedrock cosmogenic nuclide data provide local spot measurements of erosion rates that may be difficult to extrapolate spatially; hence, the measurement of spatially averaged erosion rates from river sediments (averaging over the upstream catchment) is commonly used to provide a synoptic view of regional erosion rates (von Blanckenburg 2006). In the Himalaya, such studies have shown a strong correlation of catchment-averaged erosion rates with morphology (in particular the steepness index; Godard et al. 2014; Scherler et al. 2014). The inferred erosion rates have thus been interpreted as being directly controlled by tectonics (Godard et al. 2014; Scherler et al. 2014; Le Roux-Mallouf et al. 2015), but they can be decoupled from longer-term rates (as deduced by thermochronology) and influenced by other factors, such as glacial erosion (Portenga et al. 2015; Abrahami et al. 2016).

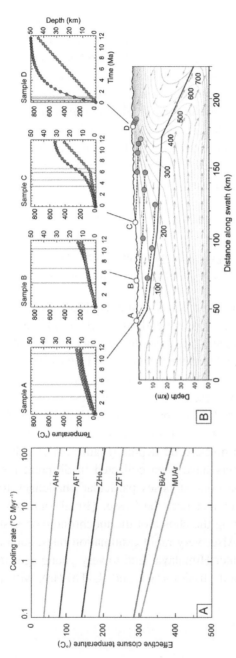

**Figure 6.4.** *Thermochronologic systems and their interpretation using thermal-kinematic modeling. (A) Effective closure temperatures for generally used thermochronologic systems as a function of cooling rate – AHe: apatite (U-Th)/He; AFT: apatite fission-track; ZHe: zircon (U-Th)/He; ZFT: zircon fission-track; BiAr: biotite 40 Ar/39 Ar; MuAr: muscovite 40 Ar/39 Ar. Modified and recalculated from Reiners and Brandon (2006). (B) Thermal-kinematic model of over- and under-thrusting along a crustal-scale detachment with a ramp-flat-ramp geometry. The gray arrows show particle paths, and the colored isotherms show the predicted thermal structure. Paths for four rock samples now exposed at the surface (white dots) are indicated, with the location where the particle paths intersect the closure temperature for the different thermochronometers (ZFT: gray; AFT: blue; AHe: green) indicated by dots. Top panels show exhumation (gray dots) and cooling (red dots) histories for these four samples; colored lines indicate the times of closure (ages) predicted for the four different systems (same color coding as the dots in the lower plot). Modified from Coutand et al. (2014). For a color version of this figure, see www.iste.co.uk/cattin/himalaya1.zip.*

## 6.2.2.2. *Thermochronology*

Thermochronology measures cooling of rocks as they are exhumed to the surface either by tectonic or erosional denudation, on temporal and spatial scales that are of the order of million years (Myr) and kilometers (km), respectively. The basis of thermochronology lies in the competition between the generation of isotopic daughter products from radiogenic decay in minerals and the removal of these daughter products by solid-state diffusion. Diffusion is a thermally activated process that is very efficient at high temperatures, leading to open isotopic systems that do not retain their daughter products, but is much more sluggish at low temperatures, leading to quantitative retainment of the daughter isotopes. The temperature at which thermochronologic systems start retaining their daughter products is known as the closure temperature. The closure temperature increases with cooling rate (Dodson 1973; Reiners and Brandon 2006), and most thermochronologic systems are characterized by a partial retention zone, i.e. a range of temperatures over which the system gradually closes, rather than a single well-defined closure temperature.

The most commonly used low-temperature thermochronologic systems are apatite and zircon fission-track, apatite and zircon (U-Th)/He and mica $^{40}Ar/^{39}Ar$. In the fission-track system, the daughter product consists of linear damage zones in the crystal lattice due to the fission of $^{238}U$; these are repaired (anneal) through similar solid-state diffusion processes as those that drive diffusion of daughter isotopes. Characteristic nominal closure temperatures for these systems vary from <100°C (for apatite (U-Th)/He) to ~350°C (for mica $^{40}Ar/^{39}Ar$ and zircon fission-track in some circumstances), allowing to track exhumation from upper to mid-crustal depths (Figure 6.4A). Although large thermochronological datasets have been collected in different regions of the Himalaya, the Himalayan context poses particular challenges to the application of these methods. For instance, the Lesser Himalayan rocks are notoriously poor in apatite, limiting the choice of thermochronologic systems to higher-temperature systems. Also, very rapid exhumation rates, such as at the topographic front of the Greater Himalaya, lead to very young ages in the lowest-temperature systems (apatite fission-track and (U-Th)/He), with large associated relative errors.

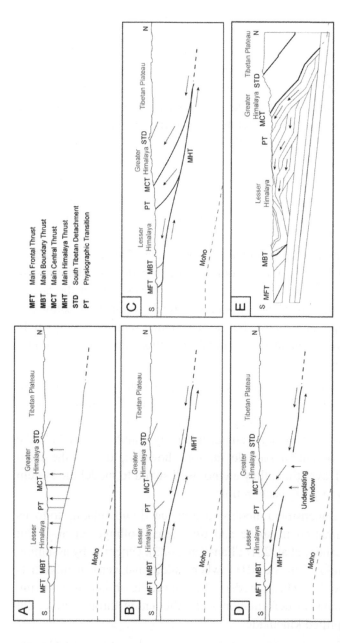

**Figure 6.5.** Conceptual sketches of different kinematics that have been used to model Himalayan exhumation patterns and their link with the geometry of the Main Himalayan Thrust (MHT): (A) one-dimensional (vertical) exhumation trajectories; (B) over- and under-thrusting over a crustal ramp in the MHT (see also thermal model in Figure 6.4B); (C) motion along surface-breaking out-of-sequence fault; (D) including underplating through an "underplating window" in the MHT; (E) structural-kinematic model based on section balancing, linked to thermal-kinematic model. See the text for discussion

### 6.2.3. *From exhumation to kinematics: thermo-kinematic models*

Advection of rocks by tectonics and surface erosion in rapidly deforming and exhuming regions such as the Himalaya leads to significant perturbation of the crustal thermal structure (e.g. Braun et al. 2006; Figure 6.4B). Moreover, in tectonic settings that are characterized by movement along crustal-scale thrusts, such as the MHT in the Himalaya, rock trajectories do not reflect simple vertical exhumation but include a strong lateral component. Finally, as outlined above, the closure temperatures of different thermochronologic systems are sensitive to cooling (and thus exhumation) rates. Thus, inferring (spatial and temporal variations in) exhumation rates from thermochronology data is complex in such settings, and a rigorous interpretation requires numerical thermal-kinematic modeling. Different modeling approaches have been developed; these are synthesized in Figure 6.5 and are discussed in order of increasing complexity as follows:

– One-dimensional thermal-kinematic models (e.g. Thiede et al. 2009; Thiede and Ehlers 2013) take into account the perturbation of isotherms by rock advection as well as the effect on closure isotherms but not the lateral motion of rocks, as they implicitly assume vertical exhumation pathways. Such models are simple and robust, but do not predict realistic rock-particle trajectories in contexts dominated by lateral advection, such as the frontal domain of the Himalaya.

– Two-dimensional models have been developed that include over- and under-thrusting along the MHT as well as potential out-of-sequence motion on other Himalayan thrusts (e.g. Whipp et al. 2007; Robert et al. 2009, 2011; Coutand et al. 2014). Such models remain relatively simple and include the first-order treatment of lateral particle motion; however, due to the fixed geometry of the MHT over time, they are not well suited to model the system more than a few million years back in time.

– Alternative two-dimensional models have included underplating to focus rock uplift and exhumation in space and time (e.g. Bollinger et al. 2006; Herman et al. 2010; Landry et al. 2016; Fan et al. 2022), generally by assuming an "underplating window" in the MHT where additional vertical motion is imposed on rock particles. These models allow more flexibility in modeling the evolution of Himalayan deformation and exhumation through time, but still feature very simplified kinematics.

– Finally, a complementary approach has been developed (e.g. McQuarrie and Ehlers 2015, 2017), which involves modeling the kinematic evolution of a fold-and-thrust belt with a structural restoration software package such as MOVE™; subsequent forward modeling then jointly predicts the structural and thermal evolution of the modeled profile. This approach attempts to model the kinematics of the evolving fold-and-thrust belt in a more realistic manner; however, it also includes numerous inherent assumptions and is highly non-unique.

The large majority of thermal-kinematic modeling studies in the Himalaya have used the *Pecube* code (Braun et al. 2012)or modifications of it. *Pecube* calculates the thermal structure and trajectories of rock particles that end up at the surface for a given tectonic scenario to predict thermal histories and thermochronologic ages that can be compared to the data. The *Pecube* code can be run in both forward and inverse mode. Inverse modeling facilitates finding robust constraints on possible scenarios and parameter ranges; this approach has been applied to the first three modeling approaches above (i.e. 1D models, 2D models with over/underthrusting only and 2D models with underplating) (Herman et al. 2010; Robert et al. 2011; Thiede and Ehlers 2013; Coutand et al. 2014; Landry et al. 2016; Fan et al. 2022). Inversions have not (yet) been applied to the joint structural–thermal evolution modeling approach, because of the large number of parameters and the complex nature of the structural restoration involved.

## 6.3. Regional case studies

### 6.3.1. *Central Himalaya – Nepal*

#### 6.3.1.1. *Kinematics of the central Nepal Himalaya*

The central Nepal Himalaya has been the subject of intense field investigations for decades and often serves as a model for the "classical" cross-sectional view of the Himalaya (Figures 6.2 and 6.6; Avouac 2003). Geodetic data from central Nepal show a major present-day uplift peak centered on the transition between the Lesser and Greater Himalaya (Jackson and Bilham 1994; Grandin et al. 2012); fluvial incision rates show a similar pattern on intemediate timescales, but with an additional peak located on the Siwalik front (Lavé and Avouac 2001). It is also in central Nepal that the steep "physiographic transition" at the front of the Greater Himalaya was first recognized and described (Hodges et al. 2001; Duncan et al. 2003; Wobus et al. 2003). Thermochronology data show very young ages, reflecting rapid exhumation in this physiographic transition zone (Burbank et al. 2003; Hodges et al. 2004; Blythe et al. 2007). South of this zone, thermochronological ages for all systems appear to increase towards the south (Figure 6.6), reflecting a decrease in exhumation rates. Some mica $^{40}$Ar/$^{39}$Ar data from the Lesser Himalaya appear incompletely reset and record pre-Himalayan ages, in particular in some detrital thermochronology samples (Wobus et al. 2006a; Ghoshal et al. 2020).

The central Nepal Himalaya is also the section of the mountain belt where the Main Himalayan Thrust (MHT) has been imaged with the best resolution (see Volume 1 – Chapter 4). The MHT was first imaged under the Tibetan plateau using active seismic sources during the INDEPTH project (Zhao et al. 1993). Subsequent studies imaged the MHT and proposed the existence of a ramp in the frontal Himalaya, during the Hi-CLIMB (Nábělek et al. 2009) and HIMNT (Schulte-Pelkum et al. 2005) acquisitions. The results of Hi-CLIMB indicate that the Moho dips gently from a depth of 40 km beneath the Indo-Gangetic plains to ~50 km beneath the Himalaya. Nábělek et al. (2009) detected an intra-crustal, strongly anisotropic layer that they interpreted as the MHT; it dips smoothly beneath central Nepal to mid-crustal depths beneath southern Tibet. The results of the HIMNT receiver-function analysis (Schulte-Pelkum et al. 2005) also provide evidence for a strongly anisotropic layer interpreted to be the MHT lying at ~8 km depth beneath the Outer (Sub) Himalaya and dipping to the north to 20 km deep just south of the Greater Himalaya (Figure 6.2b).

Diverging interpretations of the available thermochronology data in the early 2000s led to two competing models: (1) out-of-sequence thrusting on the MCT or an adjacent thrust in the physiographic transition zone, possibly climatically driven by strong erosion of the southern front of the Himalaya (Wobus et al. 2003, 2006a; Hodges et al. 2004) versus (2) focused exhumation in the topographic transition zone due to overthrusting on a crustal-scale ramp in the MHT and/or underplating and duplex development (Lavé and Avouac 2001; Bollinger et al. 2006). This debate was not solely of academic interest; the degree to which convergence is absorbed within the mountain belt or is transferred to the front along the MHT has major implications for seismic hazard, as the MHT is the nucleation zone of the major and great earthquakes threatening the Himalaya. Several thermo-kinematic modeling studies were performed to test these conflicting hypotheses (Whipp et al. 2007; Robert et al. 2009, 2011; Herman et al. 2010). The outcome of these modeling studies demonstrated that the thermochronological age pattern is not unique to a particular tectonic mechanism, and the distinction between the different models requires independent knowledge of the geometry of the involved structures. Formal inversions showed that the geometry of the MHT was not very well constrained by the thermochronologic record alone. Overall, however, the conclusion of these studies was that out-of-sequence thrusting was not necessary to explain the observed age patterns.

The Gorkha earthquake of 2015 sparked significant renewed interest in the structure of central Nepal. Geodetic inversions of co-seismic displacements were used to infer the geometry of the MHT and the potential role of shortening in the Himalayan wedge above the MHT (Elliott et al. 2016; Whipple et al. 2016); the link between structural segmentation and rupture characteristics was also studied (Hubbard et al. 2016). These developments are discussed in more detail in Volume 3 – Chapter 8, as well as in Ghoshal et al. (2020) and Dal Zilio et al. (2021).

The most recent thermal-kinematic modeling study for central Nepal (Ghoshal et al. 2020) used an approach combining structural restoration and forward modeling to test the different structural geometries proposed since the Gorkha earthquake. Ghoshal et al.'s (2020) preferred model is shown in Figure 6.6 and includes a hinterland-dipping duplex translated over a mid-crustal ramp located 110 km north of the Main Frontal Thrust, with a geometry of the MHT as inferred by Whipple et al. (2016). Ghoshal et al. (2020) also found that the temporal evolution of the ramp and duplex growth played an important role in setting the observed age distribution, an aspect that was absent in earlier models (Whipp et al. 2007; Robert et al. 2009, 2011; Herman et al. 2010). Finally, they found a slightly improved fit to the data by including some (5 km) out-of-sequence thrusting within the duplex; however, the improvement was incremental (Figure 6.6) and the fit to the data was not assessed rigorously.

### 6.3.1.2. *Western versus central Nepal Himalaya; lateral variations*

A significant change in topography occurs between central and western Nepal: in western Nepal, the mountain belt becomes much wider, with the highest peaks located about 100 km further from the mountain front than in central Nepal. The peaks are also significantly lower; no peaks over 8,000 m occur west of Dhaulagiri in central Nepal (until the Karakorum and western Himalayan syntaxis in Pakistan). Although the structure of the western Nepal Himalaya also includes a Lesser Himalayan duplex associated with a Greater Himalayan klippe (Dadeldhura; Figure 6.7), the MHT appears to be associated with much smaller ramps than in central Nepal (DeCelles et al. 2001; Robinson et al. 2006).

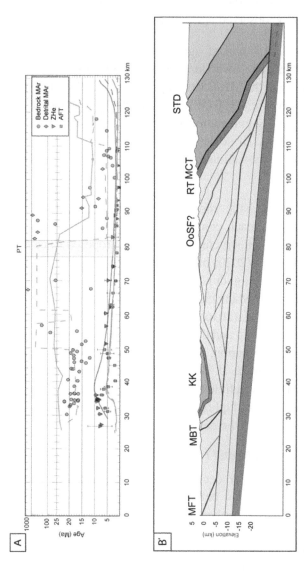

**Figure 6.6.** (A) Available thermochronology data and (B) structural cross-section along the Budi Gandhaki in central Nepal (see Figures 6.1 and 6.7 for location); abbreviations for thermochronologic systems as in Figure 6.4. Structural cross-section is the preferred model of Ghoshal et al. (2020), combining the MHT geometry proposed by Whipple et al. (2016) (purple) with a hinterland-dipping duplex in the Lesser Himalaya and potentially some out-of-sequence thrusting (red); abbreviations as in Figures 6.1 and 6.2 except KK: Kathmandu klippe; RT: Ramgarh–Munsiari Thrust. Continuous and dashed lines in panel (A) indicate predicted thermochronological ages for models without and with out-of-sequence thrusting, respectively (same color-coding as the data). Modified from Ghoshal et al. (2020). For a color version of this figure, see www.iste.co.uk/cattin/himalaya1.zip

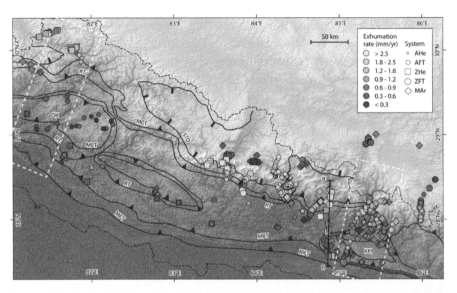

**Figure 6.7.** *Map of published thermochronology data from western and central Nepal. Symbols correspond to different thermochronometers (abbreviations as in Figure 6.4) and are colored according to average exhumation rate integrated since the time of crossing the closure temperature, calculated using the approach outlined in Schildgen et al. (2018) but adding an elevation correction as described in Willett and Brandon (2013). The dashed boxes indicate swath profiles indicated in Figure 6.8; profile line B–B' of Figures 6.2 and 6.6 is also indicated. Abbreviations of structures as in Figure 6.1, except DK: Dadeldhura klippe; KK: Kathmandu klippe; RT: Ramgarh thrust. For a color version of this figure, see www.iste.co.uk/cattin/himalaya1.zip*

Inversion of geodetic data (Jouanne et al. (2004), and Volume 3 – Chapter 7) similarly suggests only minor ramps in the MHT in western Nepal. Harvey et al. (2015) quantified the variable topography and relief between western and central Nepal and linked it to seismicity patterns and structure to propose the existence of two small ramps in the MHT of western Nepal instead of one large crustal-scale ramp in central Nepal. A recent study by van der Beek et al. (2016) reported apatite fission-track data from the Karnali River valley in western Nepal, which showed significantly older ages than in central Nepal. These authors also suggested that these differences were due to a different MHT in comparison with central Nepal, with two small ramps instead of a single large ramp (Figure 6.8). Such a geometry appears consistent with receiver-function imaging of a ~50-km long horizontal low-velocity zone interpreted as the MHT between the two ramps (Subedi et al. 2018). Recent thermo-kinematic modeling for western Nepal (Fan et al. 2022) confirms these interpretations; the thermochronologic age pattern in western Nepal was successfully reproduced by overthrusting over the southern ramp combined with underplating under the northern ramp.

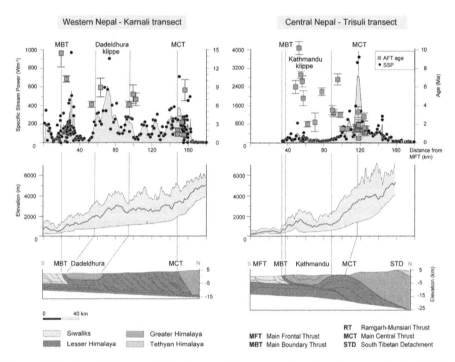

**Figure 6.8.** *Simplified geological structure (lower panels; after Jouanne et al., 2004), topographic swaths (middle panels), apatite fission-track (AFT) ages (upper panels; blue squares with error bars), and specific stream power (SSP; upper panels: dots – individual measurements; shaded area – three-point running average) projected onto transects across western (Karnali River) and central (Trisuli River) Nepal (modified from van der Beek et al. 2016). Note different scales for SSP and AFT age in the two transects. For a color version of this figure, see www.iste.co.uk/cattin/himalaya1.zip*

A major active structure has been inferred to separate western and central Nepal: the Western Nepal Fault System (Figure 6.1), which is interpreted to be a continuation of the Karakorum fault that partitions oblique convergence in the western Himalaya (Murphy and Copeland 2005; Styron et al. 2011; Murphy et al. 2014; Silver et al. 2015). Combined, these data and interpretations imply major structural segmentation between western and central Nepal.

### 6.3.1.3. *Western Himalaya – Northwest India*

The Western Himalaya is characterized by a transition between the "classical" central Himalaya, which extends from Nepal across the Kumaun and Garhwal Himalaya to the Sutlej River Valley in the west, and the Zanskar and Kashmir Himalaya that are characterized by much more oblique convergence (Figure 6.9; Kundu et al. 2014; Nennewitz et al. 2018). Similar to central Nepal, the Garhwal and Kumaun Himalaya are characterized by two peaks in rock uplift: a narrow one within the Siwaliks and a broader one underlying a prominent topographic transition ~100 km north of MFT from the Lesser to the Greater Himalaya (Thiede et al. 2004; Vannay et al. 2004; Webb et al. 2011; Colleps et al. 2019).

Present day convergence rates in the Kumaun and Garhwal Himalaya are between ~18 and ~20 mm/yr (Stevens and Avouac 2015; Yadav et al. 2019). West of the Beas River, orogen-perpendicular convergence rates drop to ~13 mm/yr (Banerjee and Bürgmann 2002; Schiffman et al. 2013; Kundu et al. 2014). There, the MBT and MCT are exposed within a few kilometers of each other (Figure 6.9), along a single topographic front that forms the hanging wall of these two main thrusts; the classical Lesser Himalaya with its intermediate topographic elevations (peaks at ~3,000 m) is missing. The topographic pattern has been interpreted as resulting from the MBT forming a deep-seated ramp merging with the MHT at depth (Thiede et al. 2017); a mid-crustal ramp further north along the MHT has not been recognized (Deeken et al. 2011). The westernmost evidence for a ramp in the MHT and/or Lesser Himalayan-duplex formation is recognized within the Kullu–Larji–Rampur and Kishtwar tectonic windows (Gavillot et al. 2018; Stübner et al. 2018, Figure 6.9). Farther west, the Quaternary Kashmir Basin is exposed within the Tethyan Himalaya (Figure 6.1), and is a unique feature within the Himalayan range. The Ramgarh–Munsiari thrust, which has been recognized as an additional orogen-scale shear zone in the proximal footwall of the MCT in the central Himalaya (Pearson and DeCelles 2005; Robinson and Pearson 2013), also terminates laterally within the Kullu–Larji–Rampur window (Stübner et al. 2018).

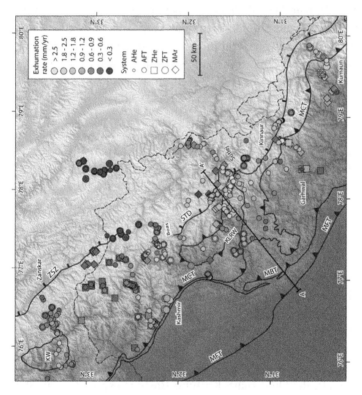

**Figure 6.9.** *Map of published thermochronology data from the north-western, Kinnaur, Garhwal and Kumaun Himalaya, NW India. Symbols correspond to different thermochronometers (abbreviations as in Figure 6.4) and are colored and calculated as in Figure 6.7. Profile line A–A' of Figure 6.2 is also indicated. Abbreviations of structures as in Figure 6.1, except KW: Kishtwar window; KLRW: Kullu–Larji–Rampur window; ZSZ: Zanskar shear zone. For a color version of this figure, see www.iste.co.uk/ cattin/himalaya1.zip*

A few seismic sections across the Western Himalaya have been reported (Rai et al. 2006; Caldwell et al. 2013; Gilligan et al. 2015; Hazarika et al. 2017; Priestley et al. 2019). Although these studies effectively constrain the depth of the Moho and its variation, the delineation of structures within the Himalayan wedge is more ambiguous. This is primarily due to the relatively low number of seismic stations along each profile, which does not allow good subsurface spatial resolution. Also, along the arc, and in some cases along a single profile, there is variation in the polarity of the impedance contrast of Ps/P amplitude on the MHT. The best-constrained mid-crustal ramp structure in the MHT has been recognized in the Garhwal Himalaya (Caldwell et al. 2013). Balanced cross-sections in the Northwest Himalaya place the ramp beneath the Munsiari Thrust (Webb et al. 2011). Both the seismic section of Caldwell et al. (2013) and structural sections by Célérier et al. (2009) recognize a continuous ramp running up to the Munsiari Thrust, which would thus be able to accommodate out-of-sequence faulting.

Late Miocene–Pliocene rapid exhumation has been linked to basal accretion beneath the Himalayan thrust belt and duplex formation in the Lesser Himalayan sequence. The late Miocene–Pliocene rapid exhumation episodes of the Lesser Himalayan Duplex coincide approximately with late Miocene to Pliocene ages of activity on the Munsiari Thrust within the Kullu–Larji–Rampur window as well as in the Sutlej section (Vannay et al. 2004; Caddick et al. 2007; Stübner et al. 2018).

Previous studies documented relatively low middle-late Miocene exhumation rates (<0.3 mm/yr; Thiede et al. 2004, 2009; Vannay et al. 2004) and attributed these to a sub-horizontal MHT that did not drive significant rock uplift and erosion in the footwall of the MCT after activity of this crustal shear zone declined around 16 Ma (Vannay et al. 2004; Stübner et al. 2018). During the late Miocene, the MHT geometry must have started resembling the present-day geometry, with the development of a ramp in the previously gently dipping MHT (Mercier et al. 2017), out-of-sequence thrusting on the Munsiari Thrust and/or duplex formation (Célérier et al. 2009; Stübner et al. 2018).

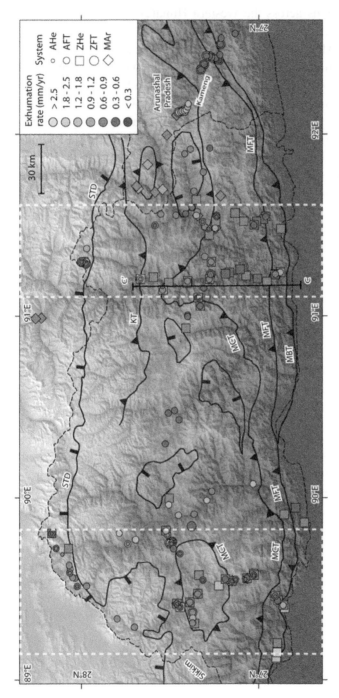

**Figure 6.10.** *Map of published thermochronology data from Bhutan and western Arunachal Pradesh (eastern Himalaya). Symbols correspond to different thermochronometers (abbreviations as in Figure 6.4) and are colored and calculated as in Figure 6.7. Dashed boxes indicate swath profiles indicated in Figure 6.11; profile line C-C' of Figure 6.2 is also indicated. Abbreviations of structures as in Figure 6.1. For a color version of this figure, see www.iste.co.uk/cattin/himalaya1.zip*

## 6.3.1.4. *Eastern Himalaya – Sikkim, Bhutan, Arunachal Pradesh*

Another major lateral topographic transition occurs in the Sikkim Himalaya, where the characteristic stepped topography of central and eastern Nepal changes to a more linear increase in topography in Bhutan (Duncan et al. 2003). Sikkim also lies at the intersection of major transverse structures, with the Yadong Rift of Southern Tibet to the north and a major basement ridge on the underthrusting Indian plate, the Munger–Saharsa ridge, to the south (Gahalaut and Kundu 2012, Figure 6.1). The Dhubri–Chungthang fault zone (Figure 6.1), a major northwest-southeast striking seismically active strike-slip fault zone connecting Sikkim to the Shillong plateau, has been suggested to control the segmentation between eastern Nepal and Bhutan (Diehl et al. 2017).

Acton et al. (2011) presented a seismic image of the lithosphere across Sikkim. Although the station spacing is wide, they clearly imaged the Moho as well as a mid-crustal low-velocity zone inferred to be the MHT. This structure contains an inferred ramp with a dip of about 15° at a distance of ~100 km from the MFT, below the surface expression of the MCT.

Thermochronology data from Sikkim show that the most rapid exhumation occurs within the Lesser Himalayan duplex exposed in the Tista half window (Abrahami et al. 2016; Landry et al. 2016). Thermal-kinematic modeling by Landry et al. (2016) showed that the thermochronology data are best fit by a model including active duplexing in the Tista half window. In contrast to observations in Nepal and western India (Godard et al. 2014; Scherler et al. 2014), short-term erosion rates from cosmogenic-nuclide data do not mirror the longer-term thermochronologically derived rates in Sikkim, but rather show the highest erosion rates in the Greater Himalaya to the north, possibly in response to localized glacial erosion (Abrahami et al. 2016).

The Bhutan Himalaya shows the archetypal linearly increasing topography that was first described there (Duncan et al. 2003). Thermochronologic ages in Bhutan tend to be older (with associated slower exhumation) than in central Nepal, for similar structural positions (Grujic et al. 2006). This difference was initially attributed to climatic variations, considering that Bhutan lies within the rain-shadow of the Shillong Plateau (Figure 6.1; Grujic et al. 2006). Another possibility is that convergence rates decelerated in the Bhutan Himalaya in the late Miocene, due to transfer of shortening to the Shillong

Plateau (Clark and Bilham 2008). Robert et al. (2011) suggested that the apparent slower exhumation rates in Bhutan were related to lateral variations in the geometry of the MHT, and used thermal-kinematic modeling to suggest a planar MHT (without a ramp) below Bhutan. The MHT geometry remained not very well constrained, however. The kinematics were significantly refined by Coutand et al. (2014), who added numerous datapoints and modeled two cross-sections, one in western and one in eastern Bhutan (Figures 6.10 and 6.11). Their modeling suggested a deep and steep ramp corresponding to the MBT/MFT (the Siwaliks being very narrow in Bhutan) and a nearly horizontal MHT extending 50-80 km northward before dipping northward at 10–15°. Coutand et al. (2014) also showed that while the western Bhutan cross-section could be modeled using a constant overthrusting velocity, fitting the eastern Bhutan data required a 40–50% decrease in overthrusting rate at 5–6 Ma, which they attributed to the outward propagation of deformation to the Shillong Plateau.

The geometry of the MHT beneath western Bhutan inferred by Coutand et al. (2014) is broadly consistent with geophysical data. Singer et al. (2017) presented a seismic image of the western Bhutan Himalaya, which showed the MHT as a sub-horizontal zone between 12 and 16 km depth for $\sim$100 km, with a $\sim$18° northward dipping mid-crustal ramp to the north (Figure 6.11). However, significant variation appears along the MHT in terms of material properties as seen through impedance contrast. A contrasting geometry was inferred by Le Roux-Mallouf et al. (2015) from a joint inversion of GPS data and millennial denudation rates from detrital cosmogenic-nuclide data; they suggested a much steeper mid-crustal ramp located nearly 100 km further north (Figure 6.11). However, this inferred geometry hinges on the interpretation that the cosmogenic-nuclide-derived denudation rates are controlled primarily by the kinematics of overthrusting, which was shown not to be the case in neighboring Sikkim (Abrahami et al. 2016).

The approach linking structural restoration and forward thermal-kinematic modeling was originally developed for eastern Bhutan (McQuarrie and Ehlers 2015, 2017), based on balanced cross-sections proposed by Long et al. (2011). These cross-sections suggested a geometry of the MHT that was fairly consistent with that inferred by Coutand et al. (2014) for eastern Bhutan (Figure 6.11), although the geological cross-sections include two small ramps in the MHT that are below the resolution of the inversion by Coutand et al. (2014), and do not include a mid-crustal ramp to the north (Figures 6.2C and 6.11). Regions of relatively young thermochronologic ages were linked to two structural duplexes in these cross-sections (Long et al. 2012; McQuarrie and Ehlers 2015; Gilmore et al. 2018): a lower Lesser Himalayan duplex in the north and an upper Lesser Himalayan duplex in the south (Figure 6.2C). Fitting the ages with these models also required considerable variability in shortening rates over time; in particular, the models require extremely rapid shortening rates of 45–75 mm/yr (i.e. well above the Himalayan convergence rate of ~20 mm/yr) during building of the upper Lesser Himalayan duplex from 13–11 to ~8 Ma (Long et al. 2012; McQuarrie and Ehlers 2015; Gilmore et al. 2018). Shortening rates would have dropped to ~7 mm/yr after ~8 Ma, consistent with the findings of Coutand et al. (2014). However, these modeling studies did not attempt a rigorous assessment of the resolution or uniqueness of their outcomes.

Moving further east, along the Kameng River section of western Arunachal Pradesh, apatite fission-track ages become younger again, reflecting faster exhumation, in particular within the Greater Himalaya where most apatite fission-track ages are <2 Ma (Figure 6.10; Adlakha et al. 2013). Adlakha et al. (2013) used simple thermal-kinematic models to infer a tectonic control on these rates, interpreting them as reflecting either a deep ramp in the MHT or out-of-sequence thrusting. A balanced cross-section (DeCelles et al. 2016) shows a duplex in the Greater Himalaya and a series of imbricate zones in the Lesser Himalaya, with km-scale ramps in the MHT stepping down to the north. Through combined structural and thermal-kinematic modeling, Braza and McQuarrie (2022) recently proposed a slight modification to this cross-section and a history of shortening rates that is somewhat similar to that inferred for Bhutan, with a peak in shortening rates that is, however, somewhat lower (25–35 mm/yr) and occurs somewhat later (between 10 and 5–7 Ma) than in Bhutan.

Singh et al. (2021) recently presented a comprehensive seismic image of the Arunachal Himalaya along several profiles, using data from 32 broadband stations. However, station spacing in each profile is sparse. They inferred a less complex crust beneath the Arunachal Himalaya in comparison to that in the Nepal and Sikkim Himalaya. Interestingly, they did not find any prominent signature of a mid-crustal ramp as observed in the Nepal Himalaya.

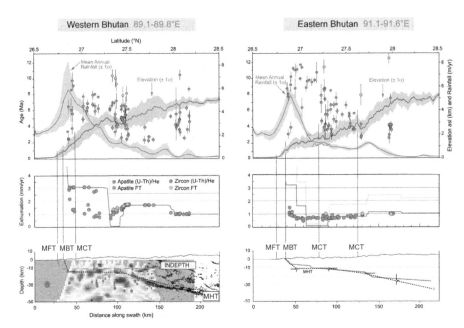

**Figure 6.11.** *Transects across western and eastern Bhutan showing – upper panels: thermochronological ages (colored circles), topography (black/gray) and precipitation (blue; bold lines show average; shading shows range of values); middle panels: average long-term exhumation rates since the time of closure indicated by the different thermochronological systems and modeled instantaneous uplift/exhumation rate from the thermal-kinematic model (red line; for eastern Bhutan both the current rate and the rate prior to 6 Ma (yellow) is shown); lower panels: best-fit geometry of the MHT inferred from modeling the thermochronology data. For western Bhutan, this geometry is compared to the seismic structure of the crust (Ps/P amplitude in red-white-blue color scale; seismic reflections from INDEPTH experiment indicated in black, from Singer et al. (2017); as well as the geometry inferred from modeling short-term erosion rates from cosmogenic-nuclide data (green dashed line; Le Roux-Mallouf et al. 2015). For eastern Bhutan, this geometry is compared to the inferred geometry of the MHT from cross-section balancing (green line; Long et al. 2011; McQuarrie and Ehlers 2015, Figure 6.2). Modified from Coutand et al. (2014). For a color version of this figure, see www.iste.co.uk/cattin/himalaya1.zip*

## 6.4. Discussion

In the following, we discuss the findings outlined above for different regions of the Himalaya within the context of a few specific questions, which link back to the general questions outlined in the introduction:

– How do the seismic, geomorphic and thermochronologic constraints on MHT geometry compare and complement each other, given the different timescales to which they pertain?

– What is the nature of the ramps in the MHT? Are they stationary or transient, and what processes control their location and size?

– Is there evidence for out-of-sequence thrusting?

– What controls the lateral segmentation of the MHT?

### 6.4.1. *Constraints on MHT geometry and kinematics at different timescales*

Geophysical data provide a snapshot of the architecture of the Himalayan wedge, whereas the geomorphologic and thermochronologic datasets provide insights on kinematics and average exhumation rates on respective timescales of $10^5 - 10^6$ and $10^6 - 10^7$ years. Geomorphic relief and channel steepness, together with channel-width measurements, help to constrain the active deformation – which should have been ongoing for some time – under the assumption that the topography has reached steady state. However, in some parts of the Himalaya, such as Western Nepal and Bhutan, uplifted relict landscapes have been recognized and document changes in deformation patterns in last few million years (Harvey et al. 2015; Adams et al. 2016).

In contrast, structural cross-sections show cumulative deformation over the timescale of building the Himalayan wedge (typically some 20 Myr) and infer the décollement geometry at depth from surface observations. Constructing these cross-sections also requires some simplifications; for instance, small-scale folding and penetrative strain is neglected. Thus, thrust sheets are treated as rigid bodies that were translated by discrete structures, and distributed strain that precedes or concurs with displacement is not incorporated. Also, all deformation is assumed to occur within the plane of the section; while this may not be a major issue in the relatively cylindrical setting of the Himalaya, out-of-section deformation could be important in transition regions between segments.

As indicated above in the regional descriptions, whereas the seismic data can generally be used to reliably map crustal thickness variations along the different sections, obtaining a high-resolution image of the MHT is much harder, due to both limited station coverage and the spatial variations in seismic expression of the MHT. Thus, all methods of characterizing the MHT have their own limitations, and a full picture will only emerge by a carefully considered combination of all approaches.

The interpretation of the kinematics of tectonic transport along the MHT similarly depends on the timescale considered. Orogenic wedges, like the Himalaya, grow primarily by frontal and basal accretion of crustal rocks, each of which lead to distinct rock-uplift patterns (Gutscher et al. 1996; Naylor and Sinclair 2007; Mercier et al. 2017; Dal Zilio et al. 2020b; Mandal et al. 2021). Instantaneous to short-term ($10^3 - 10^4$ yr) deformation can be modeled satisfactorily by passive motion over a stable ramp-flat-ramp system (Cattin and Avouac 2000; Lavé and Avouac 2001; Grandin et al. 2012; Scherler et al. 2014; Dal Zilio et al. 2021), i.e. by frontal accretion. In contrast, balanced cross-sections imply a complex history of ramp initiation, migration and abandonment accompanying the development of one or more duplexes or imbricates, and thus alternating periods of frontal and basal accretion. In particular, basal accretion and underplating is required to explain the exposure of metamorphic Lesser Himalayan units in the core of the orogen and the formation of tectonic klippen at more frontal parts of the wedge.

Although early thermal-kinematic models developed to explain the patterns of thermochronologic ages across the orogen (Whipp et al. 2007; Robert et al. 2009, 2011; Coutand et al. 2014) modeled the kinematics by simple overthrusting over a stable ramp, it is clear that such kinematics are insufficient to explain the evolution of the mountain belt over million-year timescales. Both theoretical (Naylor and Sinclair 2007; Mercier et al. 2017) and recent observational (Mandal et al. 2021) studies suggest that the Himalayan wedge may be experiencing cycles of frontal versus basal accretion on time spans of a few million years. Each cycle would then lead to migration of the active ramp in the MHT. An accretion cycle on such timescales would also be consistent with regional evidence for recent uplift of relict low-relief landscapes (Harvey et al. 2015; Adams et al. 2016).

### 6.4.2. *Nature and evolution of ramps on the MHT*

The above discussion on the geometry and evolution of the MHT leads to questions about the nature and significance of the ramps that have been identified in this structure. These questions are answered differently depending on the methods employed to study the MHT and the associated timescales. A distinction can be made between the northernmost deep crustal-scale ramp that has been evidenced in different parts of the Himalaya (most clearly in central Nepal) and more southerly ramps occurring on the MHT below the Lesser Himalaya.

In the central Himalaya, a major crustal-scale ramp in the MHT is required to link the relatively shallow detachment imaged by seismic cross-sections through the Nepal Himalaya (Schulte-Pelkum et al. 2005; Nábělek et al. 2009) to the much deeper one imaged beneath southern Tibet (Zhao et al. 1993). The top of this ramp has been shown to correlate with a band of micro-seismicity outlining the northward end of the seismically locked part of the MHT, as well as with topography >3,500 m (Stevens and Avouac 2015). This ramp has been argued to correspond to the brittle–ductile transition (Jouanne et al. 2004, 2017; Dal Zilio et al. 2021); steepening of the detachment and increase of the wedge-taper angle is consistent with predictions of critical-wedge models that are extended to include ductile deformation (Williams et al. 1994; Carena et al. 2002). Southward-younging mica $^{40}Ar/^{39}Ar$ ages above the ramp indicate it should have been active throughout the Miocene (e.g. Figure 6.6; Ghoshal et al. 2020). This ramp is not generally included in balanced cross-sections that mostly limit themselves to the Lesser Himalayan sequence above the southern flat in the MHT.

The published balanced cross-sections all show down-stepping ramps in the Lesser Himalayan units, detaching these from the underthrusting Indian plate and underplating them to the Himalayan wedge, forming a stack of Lesser Himalayan thrust nappes referred to as the Lesser Himalayan Duplex. These reconstructions also document that the ramp-flat structures are not a permanent structural feature over the time of its evolution, but rather migrate throughout Neogene times (Colleps et al. 2019), linked to the accretion cycles described above (Mercier et al. 2017; Mandal et al. 2021).

### 6.4.3. *Evidence for out-of-sequence thrusting?*

Pliocene–Recent out-of-sequence thrusting on the MCT or associated thrusts just below it has been inferred from young geo- and thermochronologic ages from low- (apatite fission-track; Hodges et al. 2004), intermediate- (mica $^{40}Ar/^{39}Ar$; Wobus et al. 2003, 2006a) as well as high-temperature (monazite and zircon-rim U-Pb; Catlos et al. 2001; Braden et al. 2018) systems. Such ages require rapid recent exhumation from mid-crustal levels and are found in rocks exposed in the hanging wall of the Lesser Himalayan Duplex, forming the southern topographic front of the Greater Himalaya. The suggestion of out-of-sequence thrusting has led to significant debate, with other authors arguing that such young ages (at least in the low- and intermediate-temperature systems) could result from rapid exhumation either over a mid-crustal ramp or in the hanging wall of a duplex (Bollinger et al. 2006; Robert et al. 2009; Herman et al. 2010).

With hindsight, it appears that the debate on out-of-sequence thrusting in the central Himalaya may have been overly polarized, as it was somewhat convolved (and confused) with a coeval debate on "channel-flow" models for Himalayan tectonics (Beaumont et al. 2001; Hodges et al. 2001; Mukherjee 2015). Some degree of out-of-sequence thrusting is expected in fold-and-thrust belts that are subject to erosion and frontal-basal accretion cycles (Hardy et al. 1998; Naylor and Sinclair 2007). Many examples of Quaternary out-of-sequence thrusting have been reported from the Siwalik fold-and-thrust belt (for a review see Mukherjee 2015), and balanced cross-section reconstructions also suggest some out-of-sequence activity on longer timescales (Robinson et al. 2006; Long et al. 2011; Webb et al. 2011; McQuarrie and Ehlers 2015). Co-seismic deformation associated with the 2015 Gorkha earthquake has been interpreted purely as reflecting slip on the MHT (Elliott et al. 2016) or as requiring active shortening (and thus, out-of-sequence thrusting) in the overlying wedge (Whipple et al. 2016).

The above discussion suggests that some degree of out-of-sequence thrusting may be expected within the Himalayan wedge. However, it appears that most thermochronological datasets do not have the resolution required to constrain the amount of, or even necessity for, out-of-sequence thrusting (Whipp et al. 2007; Herman et al. 2010; Stübner et al. 2018; Ghoshal et al. 2020). An exception to this general pattern may be the Dauladhar range in the northwest Himalaya, where modeling thermochronology data suggests sustained out-of-sequence activity on the MBT (Thiede et al. 2017). What does seem clear, however, is that any potential out-of-sequence thrusting activity would be tectonically controlled; as yet, no unequivocal evidence for a climatic control on out-of-sequence reactivation of structures has been presented.

### 6.4.4. *Lateral segmentation of the MHT*

From the above discussion of topographic and thermochronologic-age patterns, we can discern five or six main segments in the Himalayan arc (Figure 6.1): the north-western Himalaya, the Kinnaur–Garhwal–Kumaun Himalaya, western Nepal, central and eastern Nepal, Bhutan and potentially the Arunachal Pradesh Himalaya. Within these main segments, smaller-scale segmentation has been argued for on the basis of seismicity and co-seismic slip patterns (Hubbard et al. 2016; Hubbard et al. 2021). The spatial extent and temporal resolution of the currently available thermochronology data do not allow mapping this smaller-scale segmentation, with the exception of documented lateral variations in structure and kinematics between western and eastern Bhutan (Coutand et al. 2014, Figure 6.11) or between the northwest and Garhwal-Kumaun Himalaya (Eugster et al. 2018).

What controls this segmentation is still not completely clear, but several authors have pointed to pre-existing basement ridges in the underthrusting Indian plate (Valdiya 1976; Gahalaut and Kundu 2012; Dal Zilio et al. 2021; Hubbard et al. 2021, Figure 6.1). These basement ridges have been proposed to control the rupture of major and great earthquakes by acting as low-friction barriers to the incoming rupture front (Gahalaut and Kundu 2012). Although some evidence of reduced coupling indicating the presence of such barriers along the Himalayan arc has been proposed on the basis of the available GPS measurements (Dal Zilio et al. 2020a), the data are currently too limited to unequivocally demonstrate such behaviour.

A link between these basement ridges in the underthrusting Indian plate and transverse structures on the overriding plate has also been proposed (Valdiya 1976; Godin and Harris 2014; Hubbard et al. 2021). These transverse structures include the southern Tibetan rifts as well as strike-slip faults that accommodate lateral variation in obliquity of the collision and/or deformation partitioning into the foreland. The central/eastern Nepal segment is clearly delimited by two such structures, the Western Nepal Fault System to the west and the Dhubri–Chungthang fault zone to the east (Figure 6.1; Murphy et al. 2014; Silver et al. 2015; Diehl et al. 2017). These structures also roughly line up with two basement ridges on the Indian plate, the Faizabad and Munger Saharsa ridges, respectively (Figure 6.1). A third ridge, the Delhi–Hardwar ridge, may control the transition between the north-western Himalaya and Kinnaur–Garhwal–Kumaun Himalaya segments. The other segment boundaries lack a clear link to structures on either the overriding or underthrusting plate, however.

Since the ramps and flats underlying the Lesser Himalaya appear to be controlled by different potential detachment horizons within the Lesser Himalayan stratigraphy, lateral variations in the position and height of these ramps could also reflect lateral variations in the Proterozoic stratigraphic architecture of the Lesser Himalaya (Robinson et al. 2021), which could also be linked to the observed structures in the underthrusting Indian basement. Finally, it has also been suggested that the observed lateral variations in kinematics simply reflect different stages in the frontal/basal accretion cycle and the differences in structure could be linked to the rheology of the overriding plate (Duncan et al. 2003; Mercier et al. 2017).

## 6.5. Conclusion

Thermochronology data have been instrumental in constraining the longer-term kinematics of the Himalayan belt, in particular in defining the locations of ramps and/or underplating windows in the MHT, in elucidating the structure and kinematic evolution of the fold-thrust belt, and in mapping out lateral variations in these characteristics that define structural segmentation. However, on their own, these data lack the resolution required to unequivocally constrain the structure and kinematics and they need to be combined with complementary datasets. The picture that currently emerges is one in which the Himalayan belt is segmented into five or six major segments, with several potential sub-segments, the boundaries of which are at least partly controlled by pre-existing structures in both the underthrusting and overriding plates. The kinematics of the Himalayan fold-thrust belt conforms to a cycle of frontal versus basal accretion, leading to underplating and development of duplex structures. Lateral spatial variations in observed kinematics partly correspond to differences in where the particular segment is within such an accretion cycle, which could last a few million years to complete.

Significant progress has been made in the last two decades in describing and understanding the kinematics of building the Himalayan fold-thrust belt, with major implications for our understanding of the dynamics of orogenic wedges, potential tectonic-climate interactions and seismic hazard. Continued progress will require further development and comparison of complementary methods and datasets, including active deformation, seismicity, seismic imagery, geomorphology, structural geology and thermochronology. These methods all paint an image of Himalayan kinematics that is linked to a specific temporal and spatial scale; the challenge is to distil from them an overarching image that is self-consistent at all scales. Recently developed numerical models bridging the seismic to geodynamic timescales (van Dinther et al. 2013; Dal Zilio et al. 2018) could prove extremely useful in developing such a self-consistent image.

## 6.6. References

Abrahami, R., van der Beek, P., Huyghe, P., Hardwick, E., Carcaillet, J. (2016). Decoupling of long-term exhumation and short-term erosion rates in the Sikkim Himalaya. *Earth and Planetary Science Letters*, 433(C), 76–88.

Acton, C.E., Priestley, K., Mitra, S., Gaur, V.K. (2011). Crustal structure of the Darjeeling-Sikkim Himalaya and southern Tibet. *Geophysical Journal International*, 184(2), 829–852.

Adams, B.A., Whipple, K.X., Hodges, K.V., Heimsath, A.M. (2016). In-situ development of high-elevation, low-relief landscapes via duplex deformation in the Eastern Himalayan hinterland, Bhutan. *Journal of Geophysical Research*, 121(2), 294–319.

Adlakha, V., Lang, K.A., Patel, R.C., Lal, N., Huntington, K.W. (2013). Rapid long-term erosion in the rain shadow of the Shillong Plateau, Eastern Himalaya. *Tectonophysics*, 582(C), 76–83.

Andermann, C., Longuevergne, L., Bonnet, S., Crave, A., Davy, P., Gloaguen, R. (2012). Impact of transient groundwater storage on the discharge of Himalayan rivers. *Nature Geoscience*, 5(2), 127–132.

Argand, E. (1924). La tectonique de l'Asie. In *Congrès géologique international (XIIIe session)*, Brussels, 1992, 171–372.

Avouac, J.P. (2003). Mountain building, erosion, and the seismic cycle in the Nepal Himalaya. *Advances in Geophysics*, 46, 1–80.

Banerjee, P. and Bürgmann, R. (2002). Convergence across the northwest Himalaya from GPS measurements. *Geophysical Research Letters*, 29(13), 1652.

Beaumont, C., Jamieson, R.A., Nguyen, M.H., Lee, B. (2001). Himalayan tectonics explained by extrusion of a low-viscosity crustal channel coupled to focused surface denudation. *Nature*, 414(6865), 738–742.

van der Beek, P., Litty, C., Baudin, M., Mercier, J., Robert, X., Hardwick, E. (2016). Contrasting tectonically driven exhumation and incision patterns, western versus central Nepal Himalaya. *Geology*, 44(4), 327–330.

Bilham, R. (2019). Himalayan earthquakes: A review of historical seismicity and early 21st century slip potential. *Geological Society, London, Special Publications*, 483, 423–482.

Bilham, R., Gaur, V.K., Molnar, P. (2001). Himalayan seismic hazard. *Science*, 293, 1442–1444.

von Blanckenburg, F. (2006). The control mechanisms of erosion and weathering at basin scale from cosmogenic nuclides in river sediment. *Earth and Planetary Science Letters*, 242(3–4), 224–239.

Blythe, A.E., Burbank, D.W., Carter, A., Schmidt, K., Putkonen, J. (2007). Plio-Quaternary exhumation history of the central Nepalese Himalaya 1: Apatite and zircon fission track and apatite [U-Th]/He analyses. *Tectonics*, 26(3), TC3002.

Bollinger, L., Henry, P., Avouac, J.P. (2006). Mountain building in the Nepal Himalaya: Thermal and kinematic model. *Earth and Planetary Science Letters*, 244(1–2), 58–71.

Bookhagen, B. and Burbank, D.W. (2006). Topography, relief, and TRMM-derived rainfall variations along the Himalaya. *Geophysical Research Letters*, 33(8), L08405.

Bookhagen, B. and Burbank, D.W. (2010). Toward a complete Himalayan hydrological budget: Spatiotemporal distribution of snowmelt and rainfall and their impact on river discharge. *Journal of Geophysical Research*, 115(F3), F03019.

Braden, Z., Godin, L., Cottle, J., Yakymchuk, C. (2018). Renewed late Miocene (<8 Ma) hinterland ductile thrusting, western Nepal Himalaya. *Geology*, 46(6), 503–506.

Braun, J., van der Beek, P., Batt, G.E. (2006). *Quantitative Thermochronology: Numerical Methods for the Interpretation of Thermochronological Data*. Cambridge University Press.

Braun, J., van der Beek, P., Valla, P., Robert, X., Herman, F., Glotzbach, C., Pedersen, V., Perry, C., Simon-Labric, T., Prigent, C. (2012). Quantifying rates of landscape evolution and tectonic processes by thermochronology and numerical modeling of crustal heat transport using PECUBE. *Tectonophysics*, 524–525(C), 1–28.

Braza, M. and McQuarrie, N. (2022). Determining the tempo of exhumation in the eastern Himalaya: Part 1. Geometry, kinematics and predicted cooling ages. *Basin Research*, 34(1), 141–169.

Brookfield, M.E. (1998). The evolution of the great river systems of southern Asia during the Cenozoic India-Asia collision: Rivers draining southwards. *Geomorphology*, 22(3–4), 285–312.

Burbank, D.W., Blythe, A.E., Putkonen, J., Pratt-Sitaula, B. (2003). Decoupling of erosion and precipitation in the Himalayas. *Nature*, 426(6967), 652–655.

Caddick, M., Bickle, M., Harris, N., Holland, T., Horstwood, M., Parrish, R., Ahmad, T. (2007). Burial and exhumation history of a Lesser Himalayan schist: Recording the formation of an inverted metamorphic sequence in NW India. *Earth and Planetary Science Letters*, 264(3–4), 375–390.

Caldwell, W.B., Klemperer, S.L., Lawrence, J.F., Rai, S.S., Ashish (2013). Characterizing the Main Himalayan Thrust in the Garhwal Himalaya, India with receiver function CCP stacking. *Earth and Planetary Science Letters*, 367(C), 15–27.

Carena, S., Suppe, J., Kao, H. (2002). Active detachment of Taiwan illuminated by small earthquakes and its control of first-order topography. *Geology*, 30(10), 935–938.

Catlos, E.J., Harrison, T.M., Kohn, M.J., Grove, M., Ryerson, F.J., Manning, C.E., Upreti, B.N. (2001). Geochronologic and thermobarometric constraints on the evolution of the Main Central Thrust, central Nepal Himalaya. *Journal of Geophysical Research*, 106(B8), 16177–16204.

Cattin, R. and Avouac, J.P. (2000). Modeling mountain building and the seismic cycle in the Himalaya of Nepal. *Journal of Geophysical Research*, 105(B6), 13389–13407.

Célérier, J., Harrison, T.M., Beyssac, O., Herman, F., Dunlap, W.J., Webb, A.A.G. (2009). The Kumaun and Garwhal Lesser Himalaya, India: Part 2. Thermal and deformation histories. *Geological Society of America Bulletin*, 121(9–10), 1281–1297.

Clark, M.K. and Bilham, R. (2008). Miocene rise of the Shillong Plateau and the beginning of the end for the Eastern Himalaya. *Earth and Planetary Science Letters*, 269(3–4), 336–350.

Colleps, C.L., Stockli, D.F., McKenzie, N.R., Webb, A.A.G., Horton, B.K. (2019). Neogene kinematic evolution and exhumation of the NW India Himalaya: Zircon geo-/thermochronometric insights from the fold-thrust belt and foreland basin. *Tectonics*, 38(6), 2059–2086.

Coutand, I., Whipp, D.M., Grujic, D., Bernet, M., Fellin, M.G., Bookhagen, B., Landry, K.R., Ghalley, S.K., Duncan, C. (2014). Geometry and kinematics of the Main Himalayan Thrust and Neogene crustal exhumation in the Bhutanese Himalaya derived from inversion of multithermochronologic data. *Journal of Geophysical Research*, 119(2), 1446–1481.

Dal Zilio, L., van Dinther, Y., Gerya, T.V., Pranger, C.C. (2018). Seismic behaviour of mountain belts controlled by plate convergence rate. *Earth and Planetary Science Letters*, 482, 81–92.

Dal Zilio, L., Jolivet, R., van Dinther, Y. (2020a). Segmentation of the Main Himalayan Thrust illuminated by Bayesian inference of interseismic coupling. *Geophysical Research Letters*, 47(4), 67–10.

Dal Zilio, L., Ruh, J., Avouac, J. (2020b). Structural evolution of orogenic wedges: Interplay between erosion and weak décollements. *Tectonics*, 39(10), e2020TC006210.

Dal Zilio, L., Hetényi, G., Hubbard, J., Bollinger, L. (2021). Building the Himalaya from tectonic to earthquake scales. *Nature Reviews Earth Environment*, 2(4), 251–268.

DeCelles, P.G., Robinson, D.M., Quade, J., Ojha, T.P., Garzione, C.N., Copeland, P., Upreti, B.N. (2001). Stratigraphy, structure, and tectonic evolution of the Himalayan fold-thrust belt in western Nepal. *Tectonics*, 20(4), 487–509.

DeCelles, P.G., Carrapa, B., Gehrels, G.E., Chakraborty, T., Ghosh, P. (2016). Along-strike continuity of structure, stratigraphy, and kinematic history in the Himalayan thrust belt: The view from Northeastern India. *Tectonics*, 35(12), 2995–3027.

Deeken, A., Thiede, R.C., Sobel, E.R., Hourigan, J.K., Strecker, M.R. (2011). Exhumational variability within the Himalaya of northwest India. *Earth and Planetary Science Letters*, 305(1–2), 103–114.

Diehl, T., Singer, J., Hetényi, G., Grujic, D., Clinton, J., Giardini, D., Kissling, E., GANSSER Working Group (2017). Seismotectonics of Bhutan: Evidence for segmentation of the Eastern Himalayas and link to foreland deformation. *Earth and Planetary Science Letters*, 471, 54–64.

van Dinther, Y., Gerya, T.V., Dalguer, L.A., Mai, P.M., Morra, G., Giardini, D. (2013). The seismic cycle at subduction thrusts: Insights from seismo-thermo-mechanical models. *Journal of Geophysical Research: Solid Earth*, 118(12), 6183–6202.

Dodson, M.H. (1973). Closure temperature in cooling geochronological and petrological systems. *Contributions to Mineralogy and Petrology*, 40(3), 259–274.

Dunai, T.J. (2010). *Cosmogenic Nuclides: Principles, Concepts and Applications in the Earth Surface Sciences*. Cambridge University Press, Cambridge.

Duncan, C., Masek, J., Fielding, E. (2003). How steep are the Himalaya? Characteristics and implications of along-strike topographic variations. *Geology*, 31(1), 75–78.

Elliott, J.R., Jolivet, R., González, P.J., Avouac, J.P., Hollingsworth, J., Searle, M.P., Stevens, V.L. (2016). Himalayan megathrust geometry and relation to topography revealed by the Gorkha earthquake. *Nature Geoscience*, 9(2), 174–180.

Eugster, P., Thiede, R.C., Scherler, D., Stübner, K., Sobel, E.R., Strecker, M.R. (2018). Segmentation of the Main Himalayan Thrust revealed by low-temperature thermochronometry in the Western Indian Himalaya. *Tectonics*, 37(8), 2710–2726. https://doi.org/10.1029/2017tc004752.

Fan, S., Murphy, M.A., Whipp, D.M., Saylor, J.E., Copeland, P., Hoxey, A.K., Taylor, M.H., Stockli, D.F. (2022). Megathrust heterogeneity, crustal accretion, and a topographic embayment in Western Nepal Himalaya: Insights from the inversion of thermochronological data. *Tectonics*, 41, e2021TC007071. doi:10.1029/2021TC007071.

Gahalaut, V.K. and Kundu, B. (2012). Possible influence of subducting ridges on the Himalayan arc and on the ruptures of great and major Himalayan earthquakes. *Gondwana Research*, 21(4), 1080–1088.

Gansser, A. (1964). *Geology of the Himalayas*. Interscience Publishers, London, New York, Sydney.

Gavillot, Y., Meigs, A.J., Sousa, F.J., Stockli, D., Yule, D., Malik, M. (2018). Late Cenozoic foreland-to-hinterland low-temperature exhumation history of the Kashmir Himalaya. *Tectonics*, 37(9), 3041–3068.

Ghoshal, S., McQuarrie, N., Robinson, D.M., Adhikari, D.P., Morgan, L.E., Ehlers, T.A. (2020). Constraining central Himalayan (Nepal) fault geometry through integrated thermochronology and thermokinematic modeling. *Tectonics*, 39(9), e2020TC006399.

Gilligan, A., Priestley, K.F., Roecker, S.W., Levin, V., Rai, S.S. (2015). The crustal structure of the western Himalayas and Tibet. *Journal of Geophysical Research: Solid Earth*, 120(5), 3946–3964.

Gilmore, M.E., McQuarrie, N., Eizenhöfer, P.R., Ehlers, T.A. (2018). Testing the effects of topography, geometry, and kinematics on modeled thermochronometer cooling ages in the eastern Bhutan Himalaya. *Solid Earth*, 9(3), 599–627.

Godard, V., Bourlès, D.L., Spinabella, F., Burbank, D.W., Bookhagen, B., Fisher, G.B., Moulin, A., Léanni, L. (2014). Dominance of tectonics over climate in Himalayan denudation. *Geology*, 42(3), 243–246.

Godin, L. and Harris, L.B. (2014). Tracking basement cross-strike discontinuities in the Indian crust beneath the Himalayan orogen using gravity data – Relationship to upper crustal faults. *Geophysical Journal International*, 198(1), 198–215.

Grandin, R., Doin, M.P., Bollinger, L., Pinel-Puyssegur, B., Ducret, G., Jolivet, R., Sapkota, S.N. (2012). Long-term growth of the Himalaya inferred from interseismic InSAR measurement. *Geology*, 40(12), 1059–1062.

Grujic, D., Coutand, I., Bookhagen, B., Bonnet, S., Blythe, A., Duncan, C. (2006). Climatic forcing of erosion, landscape, and tectonics in the Bhutan Himalayas. *Geology*, 34(10), 801–14.

Grujic, D., Warren, C.J., Wooden, J.L. (2011). Rapid synconvergent exhumation of Miocene-aged lower orogenic crust in the eastern Himalaya. *Lithosphere*, 3(5), 346–366.

Gutscher, M.-A., Kukowski, N., Malavieille, J., Lallemand, S. (1996). Cyclical behavior of thrust wedges: Insights from high basal friction sandbox experiments. *Geology*, 24(2), 135–138.

Hardy, S., Duncan, C., Masek, J., Brown, D. (1998). Minimum work, fault activity and the growth of critical wedges in fold and thrust belts. *Basin Research*, 10(3), 365–373.

Harvey, J.E., Burbank, D.W., Bookhagen, B. (2015). Along-strike changes in Himalayan thrust geometry: Topographic and tectonic discontinuities in western Nepal. *Lithosphere*, 7(5), 511–518.

Hazarika, D., Wadhawan, M., Paul, A., Kumar, N., Borah, K. (2017). Geometry of the Main Himalayan Thrust and Moho beneath Satluj valley, northwest Himalaya: Constraints from receiver function analysis. *Journal of Geophysical Research: Solid Earth*, 122(4), 2929–2945.

Herman, F., Copeland, P., Avouac, J.-P., Bollinger, L., Mahéo, G., Le Fort, P., Rai, S., Foster, D., Pêcher, A., Stüwe, K., Henry, P. (2010). Exhumation, crustal deformation, and thermal structure of the Nepal Himalaya derived from the inversion of thermochronological and thermobarometric data and modeling of the topography. *Journal of Geophysical Research*, 115(B6), B06407.

Hintersberger, E., Thiede, R.C., Strecker, M.R. (2011). The role of extension during brittle deformation within the NW Indian Himalaya. *Tectonics*, 30(3), TC3012.

Hodges, K.V. (2000). Tectonics of the Himalaya and southern Tibet from two perspectives. *Geological Society of America Bulletin*, 112(3), 324–350.

Hodges, K.V., Hurtado, J.M., Whipple, K.X. (2001). Southward extrusion of Tibetan crust and its effect on Himalayan tectonics. *Tectonics*, 20(6), 799–809.

Hodges, K.V., Wobus, C., Ruhl, K., Schildgen, T., Whipple, K. (2004). Quaternary deformation, river steepening, and heavy precipitation at the front of the Higher Himalayan ranges. *Earth and Planetary Science Letters*, 220(3–4), 379–389.

Hubbard, J., Almeida, R., Foster, A., Sapkota, S.N., Bürgi, P., Tapponnier, P. (2016). Structural segmentation controlled the 2015 Mw 7.8 Gorkha earthquake rupture in Nepal. *Geology*, 44(8), 639–642.

Hubbard, M., Mukul, M., Gajurel, A.P., Ghosh, A., Srivastava, V., Giri, B., Seifert, N., Mendoza, M.M. (2021). Orogenic segmentation and its role in Himalayan mountain building. *Frontiers in Earth Science*, 9, 641666.

Jackson, M. and Bilham, R. (1994). Constraints on Himalayan deformation inferred from vertical velocity fields in Nepal and Tibet. *Journal of Geophysical Research*, 99(B7), 13897–13912.

Jouanne, F., Mugnier, J.L., Gamond, J.F., Le Fort, P., Pandey, M.R., Bollinger, L., Flouzat, M., Avouac, J.P. (2004). Current shortening across the Himalayas of Nepal. *Geophysical Journal International*, 157(1), 1–14.

Jouanne, F., Mugnier, J.-L., Sapkota, S.N., Bascou, P., Pêcher, A. (2017). Estimation of coupling along the Main Himalayan Thrust in the central Himalaya. *Journal of Asian Earth Sciences*, 133, 62–71.

Kapp, P. and DeCelles, P.G. (2019). Mesozoic-Cenozoic geological evolution of the Himalayan-Tibetan orogen and working tectonic hypotheses. *American Journal of Science*, 319, 159–254.

Kirby, E. and Whipple, K.X. (2012). Expression of active tectonics in erosional landscapes. *Journal of Structural Geology*, 44(C), 54–75.

Kundu, B., Yadav, R.K., Bali, B.S., Chowdhury, S., Gahalaut, V.K. (2014). Oblique convergence and slip partitioning in the NW Himalaya: Implications from GPS measurements. *Tectonics*, 33(10), 2013–2024.

Lague, D. (2014). The stream power river incision model: Evidence, theory and beyond. *Earth Surface Processes and Landforms*, 39(1), 38–69.

Landry, K.R., Coutand, I., Whipp, D.M., Grujic, D., Hourigan, J.K. (2016). Late Neogene tectonically driven crustal exhumation of the Sikkim Himalaya: Insights from inversion of multithermochronologic data. *Tectonics*, 35(3), 833–859.

Lavé, J. and Avouac, J.P. (2001). Fluvial incision and tectonic uplift across the Himalayas of central Nepal. *Journal of Geophysical Research*, 106(B11), 26561–26591.

Le Fort, P. (1975). Himalayas: The collided range. Present knowledge of the continental arc. *American Journal of Science*, 275-A, 1–44.

Le Roux-Mallouf, R., Godard, V., Cattin, R., Ferry, M., Gyeltshen, J., Ritz, J.-F., Drupka, D., Guillou, V., Arnold, M., Aumaître, G. (2015). Evidence for a wide and gently dipping Main Himalayan Thrust in western Bhutan. *Geophysical Research Letters*, 42(9), 3257–3265.

Long, S.P., McQuarrie, N., Tobgay, T., Grujic, D. (2011). Geometry and crustal shortening of the Himalayan fold-thrust belt, eastern and central Bhutan. *Geological Society of America Bulletin*, 123(7–8), 1427–1447.

Long, S.P., McQuarrie, N., Tobgay, T., Coutand, I., Cooper, F.J., Reiners, P.W., Wartho, J.-A., Hodges, K.V. (2012). Variable shortening rates in the eastern Himalayan thrust belt, Bhutan: Insights from multiple thermochronologic and geochronologic data sets tied to kinematic reconstructions. *Tectonics*, 31(5), TC5004.

Mandal, S.K., Robinson, D.M., Kohn, M.J., Khanal, S., Das, O. (2019). Examining the tectono-stratigraphic architecture, structural geometry, and kinematic evolution of the Himalayan fold-thrust belt, Kumaun, northwest India. *Lithosphere*, 11(4), 414–435.

Mandal, S.K., Scherler, D., Wittmann, H. (2021). Tectonic accretion controls erosional cyclicity in the Himalaya. *AGU Advances*, 2(3)2, e2021AV000487. doi: 10.1029/2021AV000539.

McQuarrie, N. and Ehlers, T.A. (2015). Influence of thrust belt geometry and shortening rate on thermochronometer cooling ages: Insights from thermokinematic and erosion modeling of the Bhutan Himalaya. *Tectonics*, 34(6), 1055–1079.

McQuarrie, N. and Ehlers, T.A. (2017). Techniques for understanding fold-and-thrust belt kinematics and thermal evolution. In *Linkages and Feedbacks in Orogenic System*, Law, R.D., Thigpen, J.R., Merschat, A.J., Stowell, H.H. (eds). Geological Society of America, Boulder, CO.

McQuarrie, N., Robinson, D., Long, S., Tobgay, T., Grujic, D., Gehrels, G., Ducea, M. (2008). Preliminary stratigraphic and structural architecture of Bhutan: Implications for the along strike architecture of the Himalayan system. *Earth and Planetary Science Letters*, 272(1–2), 105–117.

Mercier, J., Braun, J., van der Beek, P. (2017). Do along-strike tectonic variations in the Nepal Himalaya reflect different stages in the accretion cycle? Insights from numerical modeling. *Earth and Planetary Science Letters*, 472, 299–308.

Mukherjee, S. (2015). A review on out-of-sequence deformation in the Himalaya. *Geological Society, London, Special Publications*, 412(1), 67–109.

Murphy, M.A. and Copeland, P. (2005). Transtensional deformation in the central Himalaya and its role in accommodating growth of the Himalayan orogen. *Tectonics*, 24(4), TC4012.

Murphy, M.A., Taylor, M.H., Gosse, J., Silver, C., Whipp, D.M., Beaumont, C. (2014). Limit of strain partitioning in the Himalaya marked by large earthquakes in western Nepal. *Nature Geoscience*, 7, 38–42.

Nábělek, J., Hetényi, G., Vergne, J., Sapkota, S., Kafle, B., Jiang, M., Su, H., Chen, J., Huang, B.S., Hi-CLIMB Team (2009). Underplating in the Himalaya-Tibet collision zone revealed by the Hi-CLIMB experiment. *Science*, 325(5946), 1371–1374.

Naylor, M. and Sinclair, H.D. (2007). Punctuated thrust deformation in the context of doubly vergent thrust wedges: Implications for the localization of uplift and exhumation. *Geology*, 35(6), 559–562.

Nennewitz, M., Thiede, R.C., Bookhagen, B. (2018). Fault activity, tectonic segmentation, and deformation pattern of the western Himalaya on Ma timescales inferred from landscape morphology. *Lithosphere*, 10(5), 632–640.

Pearson, O.N. and DeCelles, P.G. (2005). Structural geology and regional tectonic significance of the Ramgarh thrust, Himalayan fold-thrust belt of Nepal. *Tectonics*, 24(4), TC4008.

Portenga, E.W., Bierman, P.R., Duncan, C., Corbett, L.B., Kehrwald, N.M., Rood, D.H. (2015). Erosion rates of the Bhutanese Himalaya determined using in situ-produced 10Be. *Geomorphology*, 233(C), 112–126.

Priestley, K., Ho, T., Mitra, S. (2019). The crustal structure of the Himalaya: A synthesis. *Geological Society, London, Special Publications*, 483, 483–516.

Rai, S.S., Priestley, K.F., Gaur, V.K., Mitra, S., Singh, M.P., Searle, M. (2006). Configuration of the Indian Moho beneath the NW Himalaya and Ladakh. *Geophysical Research Letters*, 33(15), L15308.

Reiners, P.W. and Brandon, M.T. (2006). Using thermochronology to understand orogenic erosion. *Annual Review of Earth and Planetary Sciences*, 34, 419–466.

Reiners, P.W., Carlson, R.W., Renne, P.R., Cooper, K.M., Granger, D.E., McLean, N.M., Schoene, B. (2017). *Geochronology and Thermochronology*. John Wiley & Sons Ltd, Hoboken, NJ.

Robert, X., van der Beek, P., Braun, J., Perry, C., Dubille, M., Mugnier, J.L. (2009). Assessing Quaternary reactivation of the Main Central Thrust zone (central Nepal Himalaya): New thermochronologic data and numerical modeling. *Geology*, 37(8), 731–734.

Robert, X., van der Beek, P., Braun, J., Perry, C., Mugnier, J.-L. (2011). Control of detachment geometry on lateral variations in exhumation rates in the Himalaya: Insights from low-temperature thermochronology and numerical modeling. *Journal of Geophysical Research*, 116(B5), B05202.

Robinson, D.M. and Pearson, O.N. (2013). Was Himalayan normal faulting triggered by initiation of the Ramgarh–Munsiari thrust and development of the Lesser Himalayan duplex? *International Journal of Earth Sciences*, 102(7), 1773–1790.

Robinson, D.M., DeCelles, P.G., Copeland, P. (2006). Tectonic evolution of the Himalayan thrust belt in western Nepal: Implications for channel flow models. *Geological Society of America Bulletin*, 118(7–8), 865–885.

Robinson, D.M., Khanal, S., Olree, E., Bhattacharya, G., Mandal, S. (2021). Controls of stratigraphic architecture on along strike cooling age patterns. *Terra Nova*, 33(2), 129–136.

Schelling, D. and Arita, K. (1991). Thrust tectonics, crustal shortening, and the structure of the far-eastern Nepal Himalaya. *Tectonics*, 10(5), 851–862.

Scherler, D., Bookhagen, B., Strecker, M.R. (2014). Tectonic control on 10Be-derived erosion rates in the Garhwal Himalaya, India. *Journal of Geophysical Research: Earth Surface*, 119, 83–105.

Schiffman, C., Bali, B.S., Szeliga, W., Bilham, R. (2013). Seismic slip deficit in the Kashmir Himalaya from GPS observations. *Geophysical Research Letters*, 40(21), 5642–5645.

Schildgen, T.F., van der Beek, P.A., Sinclair, H.D., Thiede, R.C. (2018). Spatial correlation bias in late-Cenozoic erosion histories derived from thermochronology. *Nature*, 559(7712), 89–93.

Schulte-Pelkum, V., Monsalve, G., Sheehan, A., Pandey, M.R., Sapkota, S., Bilham, R., Wu, F. (2005). Imaging the Indian subcontinent beneath the Himalaya. *Nature*, 435(7046), 1222–1225.

Seeber, L. and Gornitz, V. (1983). River profiles along the Himalayan arc as indicators of active tectonics. *Tectonophysics*, 92(4), 335–367.

Seeber, L., Armbruster, J.G., Quittmeyer, R.C. (1981). Seismicity and continental subduction in the Himalayan arc. In *Zagros Hindu Kush Himalaya Geodynamic Evolution*, Gupta, H.K. and Delany, F.M. (eds). American Geophysical Union, Washington, DC.

Silver, C., Murphy, M.A., Taylor, M.H., Gosse, J., Baltz, T. (2015). Neotectonics of the Western Nepal Fault System: Implications for Himalayan strain partitioning. *Tectonics*, 34(12), 2494–2513.

Singer, J., Kissling, E., Diehl, T., Hetényi, G. (2017). The underthrusting Indian crust and its role in collision dynamics of the Eastern Himalaya in Bhutan: Insights from receiver function imaging. *Journal of Geophysical Research: Solid Earth*, 122(2), 1152–1178.

Singh, A., Saikia, D., Kumar, M.R. (2021). Seismic imaging of the crust beneath Arunachal Himalaya. *Journal of Geophysical Research: Solid Earth*, 126(3), e2020JB020616.

Srivastava, P. and Mitra, G. (1994). Thrust geometries and deep structure of the outer and lesser Himalaya, Kumaon and Garhwal (India): Implications for evolution of the Himalayan fold-and-thrust belt. *Tectonics*, 13(1), 89–109.

Stevens, V.L. and Avouac, J.P. (2015). Interseismic coupling on the main Himalayan thrust. *Geophysical Research Letters*, 42(14), 5828–5837.

Stübner, K., Grujic, D., Dunkl, I., Thiede, R., Eugster, P. (2018). Pliocene episodic exhumation and the significance of the Munsiari thrust in the northwestern Himalaya. *Earth and Planetary Science Letters*, 481, 273–283.

Styron, R.H., Taylor, M.H., Murphy, M.A. (2011). Oblique convergence, arc-parallel extension, and the role of strike-slip faulting in the High Himalaya. *Geosphere*, 7(2), 582–596.

Subedi, S., Hetényi, G., Vergne, J., Bollinger, L., Lyon-Caen, H., Farra, V., Adhikari, L.B., Gupta, R.M. (2018). Imaging the Moho and the Main Himalayan Thrust in Western Nepal with receiver functions. *Geophysical Research Letters*, 45(24), 13,222–13,230

Thiede, R.C. and Ehlers, T.A. (2013). Large spatial and temporal variations in Himalayan denudation. *Earth and Planetary Science Letters*, 371–372(C), 278–293.

Thiede, R.C., Bookhagen, B., Arrowsmith, J.R., Sobel, E.R., Strecker, M.R. (2004). Climatic control on rapid exhumation along the Southern Himalayan Front. *Earth and Planetary Science Letters*, 222(3–4), 791–806.

Thiede, R.C., Arrowsmith, J.R., Bookhagen, B., McWilliams, M.O., Sobel, E.R., Strecker, M.R. (2005). From tectonically to erosionally controlled development of the Himalayan orogen. *Geology*, 33(8), 689–692.

Thiede, R.C., Ehlers, T.A., Bookhagen, B., Strecker, M.R. (2009). Erosional variability along the northwest Himalaya. *Journal of Geophysical Research*, 114(F1), F01015.

Thiede, R.C., Robert, X., Stübner, K., Dey, S., Faruhn, J. (2017). Sustained out-of-sequence shortening along a tectonically active segment of the Main Boundary thrust: The Dhauladhar Range in the northwestern Himalaya. *Lithosphere*, 9(5), 715–725.

Valdiya, K.S. (1976). Himalayan transverse faults and folds and their parallelism with subsurface structures of North Indian plains. *Tectonophysics*, 32(3), 353–386.

Vannay, J.-C., Grasemann, B., Rahn, M., Frank, W., Carter, A., Baudraz, V., Cosca, M. (2004). Miocene to Holocene exhumation of metamorphic crustal wedges in the NW Himalaya: Evidence for tectonic extrusion coupled to fluvial erosion. *Tectonics*, 23(1), TC1014.

Webb, A.A.G., Yin, A., Harrison, T.M., Célérier, J., Gehrels, G.E., Manning, C.E., Grove, M. (2011). Cenozoic tectonic history of the Himachal Himalaya (northwestern India) and its constraints on the formation mechanism of the Himalayan orogen. *Geosphere*, 7(4), 1013–1061.

Whipp, D.M., Ehlers, T.A., Blythe, A.E., Huntington, K.W., Hodges, K.V., Burbank, D.W. (2007). Plio-Quaternary exhumation history of the central Nepalese Himalaya 2: Thermokinematic and thermochronometer age prediction model. *Tectonics*, 26(3), TC3003.

Whipple, K.X. (2004). Bedrock rivers and the geomorphology of active orogens. *Annual Review of Earth and Planetary Sciences*, 32(1), 151–185.

Whipple, K.X., Shirzaei, M., Hodges, K.V., Arrowsmith, J.R. (2016). Active shortening within the Himalayan orogenic wedge implied by the 2015 Gorkha earthquake. *Nature Geoscience*, 9, 711–716.

Willett, S.D. and Brandon, M.T. (2013). Some analytical methods for converting thermochronometric age to erosion rate. *Geochemistry, Geophysics, Geosystems*, 14(1), 209–222.

Williams, C.A., Connors, C., Dahlen, F.A., Price, E.J., Suppe, J. (1994). Effect of the brittle-ductile transition on the topography of compressive mountain belts on Earth and Venus. *Journal of Geophysical Research: Solid Earth*, 99(B10), 19947–19974.

Wobus, C.W., Hodges, K.V., Whipple, K.X. (2003). Has focused denudation sustained active thrusting at the Himalayan topographic front? *Geology*, 31(10), 861–864.

Wobus, C.W., Heimsath, A., Whipple, K.X., Hodges, K. (2005). Active out-of-sequence thrust faulting in the central Nepalese Himalaya. *Nature*, 434(7036), 1008–1011.

Wobus, C.W., Whipple, K.X., Hodges, K.V. (2006a). Neotectonics of the central Nepalese Himalaya: Constraints from geomorphology, detrital 40Ar/39Ar thermochronology, and thermal modeling. *Tectonics*, 25(4), TC4011.

Wobus, C.W., Whipple, K.X., Kirby, E., Snyder, N., Johnson, J., Spyropolou, K., Crosby, B., Sheehan, D. (2006b). Tectonics from topography: Procedures, promise, and pitfalls. Special paper 398, Geological Society of America, 55–74.

Yadav, R.K., Gahalaut, V.K., Bansal, A.K., Sati, S.P., Catherine, J., Gautam, P., Kireet Kumar, K., Rana, N. (2019). Strong seismic coupling underneath Garhwal–Kumaun region, NW Himalaya, India. *Earth and Planetary Science Letters*, 506, 8–14.

Yin, A. and Harrison, T.M. (2000). Geologic evolution of the Himalayan–Tibetan orogen. *Annual Review of Earth and Planetary Sciences*, 28(1), 211–280.

Zhao, W., Nelson, K.D., Che, J., Quo, J., Lu, D., Wu, C., Liu, X. (1993). Deep seismic reflection evidence for continental underthrusting beneath southern Tibet. *Nature*, 366(6455), 557–559.

# PART 3

# Focus

PART 3

# 7

# Application of Near-surface Geophysical Methods for Imaging Active Faults in the Himalaya

Dowchu Drukpa[1], Stéphanie Gautier[2] and Rodolphe Cattin[2]

[1]*Earthquake and Geophysics Division, Department of Geology and Mines, Thimphu, Bhutan*

[2]*University of Montpellier, France*

## 7.1. Introduction

Near-surface geophysical investigation is a relevant tool to provide quantitative constraints of the nature and the geometry of shallow fault zones. The strategy consists of combining results coming from different geophysical methods, including electrical resistivity, seismic and gravity measurements, to obtain a detailed characterization of the first 100s of m in terms of geometry (faults, stratigraphy, buried markers), cumulative deformation on the downthrow and quantification of the physical parameters around the fault. Integrating the new constraints on fault geometry and geomorphological and regional geodetic information allows one to estimate fault slip rate and to identify potential slip partitioning between horizontal and vertical

*Himalaya, Dynamics of a Giant 1,*
coordinated by Rodolphe Cattin and Jean-Luc Epard.
© ISTE Ltd 2023.

displacement. The characterization of the seismic sources and slip partitioning has major implications in seismic hazard evaluation in the Himalaya area.

## 7.2. Near-surface geophysics

Geophysical methods have been widely used for the characterization of subsurface tectonic features (Suzuki et al. 2000; Demanet et al. 2001; Louis et al. 2002; Morandi and Ceragioli 2002; Wise et al. 2003; Nguyen et al. 2005, 2007; Kaiser et al. 2009). Depending on the scale of investigation, geophysical methods can be divided into two categories (Mussett et al. 2000): deep surveys, with penetration depth ranges from 100 m to several kilometers, which are mainly used to define regional seismotectonic models, and near-surface surveys, which image structures at shallow depth (0–200 m) such as fault systems, lithological interfaces, landslide bodies as well as cumulative deformation. The same near-surface geophysical techniques are also deployed to quantify groundwater resources, to monitor active geohazards, for archeogeophysical exploration and geotechnical site characterization (Telford et al. 1990; Reynolds 1997). The success of any geophysical investigation is dependent on several factors, including:

– the presence of strong contrasts in terms of physical properties;

– the availability of geological, geomorphological or hydrological information;

– practical aspects for the deployment such as topography or site accessibility.

Having good a priori information is critically important to choose relevant locations for the surveys and validate the geophysical investigations' outcome.

Various near-surface geophysical methods are particularly adapted to image internal structures and physical properties of fault zones, in a depth range between a few meters to a hundred meters (Demanet et al. 2001; Villani et al. 2015). In particular, geophysical data can contribute to characterizing the geometry of the fault dip angle, the thickness of quaternary layers and possible offsets at shallow depths. Those data can also help detect blind faults close to the surface and find the appropriate location for trench excavations. Despite promising results on many fault zones around the world (Nguyen

2005), near-surface geophysical investigations have been less implemented in the Himalaya.

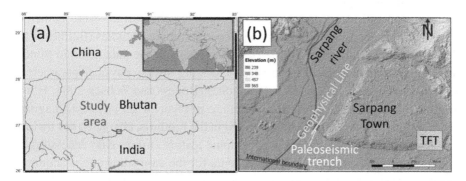

**Figure 7.1.** *(a) Location of the Kingdom of Bhutan and the study area in south Central part of Bhutan. (b) High-resolution Digital Elevation model (DEM) from satellite image of the study area showing the Topographic Frontal Thrust (TFT) fault trace, the location of the paleoseismic trench studied by Le Roux-Mallouf et al. (2016) and our geophysical profile (yellow line). For a color version of this figure, see www.iste.co.uk/cattin/himalaya1.zip*

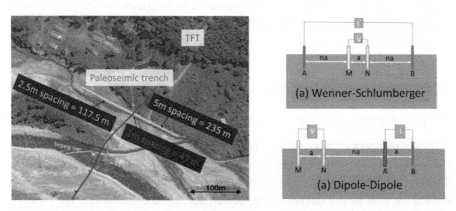

**Figure 7.2.** *Location of both the Electrical Resistivity Tomography (ERT) surveys (yellow, blue and red lines for the 5 m, 2.5 m and 1 m electrode spacing soundings, respectively) and the paleoseismic trench in Sarpang site (left) and geometric configurations used in the study (right). For a color version of this figure, see www.iste.co.uk/cattin/himalaya1.zip*

In this framework, near-surface geophysical investigations were carried out in central southern Bhutan for shallow subsurface imaging of the Topographic Frontal Thrust (Figure 7.1). At the front, a priori information of the exact location and estimated dip angle of the fault on the surface were gathered from previous geomorphological and paleoseismological studies (Berthet et al. 2014; Le Roux-Mallouf et al. 2016). Observations in the paleoseismic trench also confirmed rheological contrasts, suggesting strong physical parameters variations as usually described in fault zones (Boness and Zoback 2004; Hung et al. 2009; Jeppson et al. 2010), which are in favor of geophysical imaging. Furthermore, the case study site is located in the foothills, characterized by low elevation variations. This specific geographic feature ensures easy accessibility and feasibility of surveys. Geophysical data for the case study site is acquired along the east side of Sarpang river, where the east-west trending TFT trace was intersected by the previous paleoseismic trench study down to 1 m depth (Figure 7.1). Using various geophysical methods (Electrical resistivity tomography (ERT), seismic refraction and gravity measurements) provides images of different physical properties at different depths of investigation. All geophysical data were collected along the same N-S profile, with varying lengths of spread depending on the methods. The midpoints of the geophysical surveys were positioned at the fault location at the surface deduced from the paleoseismic study.

### 7.2.1. *Geophysical methods for fault mapping*

#### 7.2.1.1. *Electrical resistivity tomography (ERT)*

Low resistivity materials generally characterize fault zones. Hence, electrical resistivity tomography is widely used for near-surface fault imaging (Phillips and Kuckes 1983). To obtain well-resolved images close to the surface as well as the deeper depth of investigation, combining ERT profiles with different electrode spacing is recommended (Nguyen et al. 2005; Gelis et al. 2010). Among the different geometric configurations available for ERT surveys, the commonly used Wenner–Schlumberger (WS) and dipole–dipole (DD) configurations (Figure 7.2) seem to achieve a good compromise between vertical and horizontal resolution and effect of noise (Dahlin and Zhou 2004; Loke 2015).

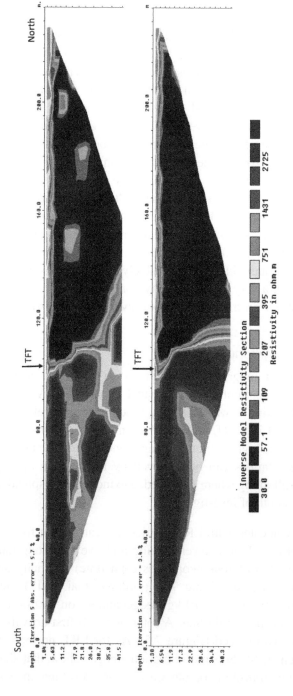

**Figure 7.3.** *2D 5m-spacing dipole–dipole ERT model (top) and Wenner–Schlumberger model (bottom) inverted with* RES2DINV *(Loke and Barker 1996). Both images represent models obtained after five iterations. The RMS corresponds to the misfit between observed and computed data. The TFT label indicates the location of fault. For a color version of this figure, see www.iste.co.uk/cattin/himalaya1.zip*

Accordingly, in south Bhutan, both WS and DD electrical resistivity soundings, with three different electrode spacings (1, 2.5 and 5 m), were carried out (Drukpa et al. 2017). The topography along the profile is relatively smooth, with variations less than 1.3 m. Therefore, no topographic correction was necessary for resistivity data analyses. The obtained ERT inverted sections (Figure 7.3) illustrate that an electrical resistivity survey is a valuable tool for characterizing faults in superficial layers from the ground surface. The different surveys provided consistent results. The WS electrical images appear to be more robust because of a greater sensitivity to both lateral and vertical variations (Nguyen et al. 2007). These resistivity images point out a major sub-vertical discontinuity, which is consistent with the prolongation of the fault towards the surface. More precisely, the fault zone is marked by high electrical resistivity contrasts ($\sim$1:100) with a nearly vertical contact down to $\sim$40 m depth. The north side shows a uniform apparent resistivity layering with a thin upper layer resistivity of 200–1,000 $\Omega \cdot m$ overlying a very low resistivity layer < 100 $\Omega \cdot m$. The south side shows relatively constant resistivity values (1,000–4,000 $\Omega \cdot m$) with a very high resistivity zone located at 5–15 m depth at the southern end of our profile. A thin upper layer of low resistivity is also observed southward.

## 7.2.1.2. *Seismic tomography*

The passive seismic tomography technique is an additional method for characterizing near-surface fault zone (Demanet et al. 2001; Villani et al. 2015) by usually estimating P-wave velocity models from travel times (see Volume 1 – Chapter 4). Like ERT, data acquisition layout design for seismic tomography should obtain high-resolution images and achieve the target depth of investigation. These targets can be achieved by maintaining appropriate receiver spacing and roll-along where ever needed to increase the spread length and, therefore deeper depth of investigation.

A seismic survey colocated with ERT profiles was carried out for the case study. It is composed of a 1 m receiver spacing and five roll-along (shift of 20 geophones and overlap of four geophones each time) to finally acquire a 103 m long seismic profile (Figure 7.4) (Drukpa et al. 2017). Seismic sources were generated by hitting a 10 kg sledgehammer on an iron plate at each geophone along each seismic line. A total of 5,760 first-arrivals travel times were manually hand-picked (Figure 7.5). As for ERT, no topographic correction was implemented. Seismic refraction images show that the TFT fault implies an abrupt transition of the travel times resulting in a strong contrast of both the ray distribution and the velocity values (Figure 7.5). The velocity model confirms the presence of a shallow interface on the northern side.

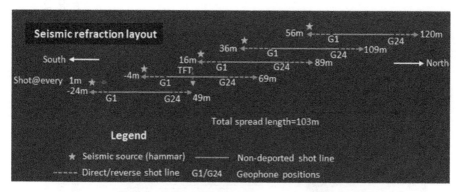

**Figure 7.4.** *Seismic data acquisition layout plan. For a color version of this figure, see www.iste.co.uk/cattin/himalaya1.zip*

### 7.2.1.3. *Microgravity*

The gravity method is a versatile geophysical technique to determine density contrasts within the Earth. Variations of the gravitational field due to the density contrasts are measured using extremely sensitive instruments to identify anomalies at depth. This technique allows determining and locating the presence of a deficit or an excess of mass in the subsurface, which corresponds to negative or positive anomalies (see Volume 1 – Chapter 5). In near-surface geophysics, microgravimetry is carried out for various investigations such as detection of karsts and voids, measurements of sediment thickness, archaeological surveys or mineral exploration (Telford et al. 1990).

In southern Bhutan, gravity measurements were recorded along the same south–north profile as the ERT and seismic lines (Drukpa et al. 2017). From the center point of the survey line positioned at the paleoseismic trench, gravity readings were acquired at every 5 m on either side of the profile covering a distance of 30 m and 105 m to the south and the north, respectively. Spatially denser gravity points were collected in the vicinity of the fault area. Using the GravProcess software (Cattin et al. 2015), network adjustment was performed and topographic effect was corrected from accurate elevation data gathered along the same profile assuming a constant density of 2,670 kg/m$^3$.

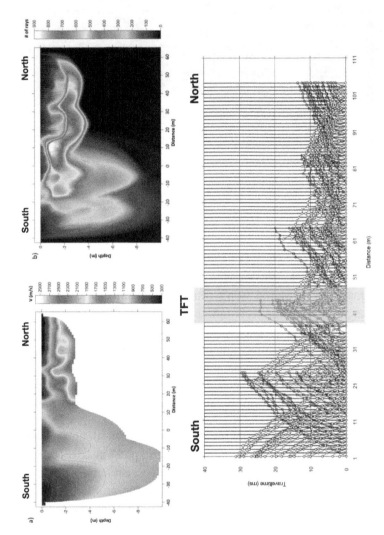

**Figure 7.5.** Top: (a) seismic tomographic refraction image showing the velocity variations of both sides of the TFT obtained using RAYFRACT software; The TFT fault trace is located at 0 m. (b) Ray coverage illustrating the resolved area. Bottom: hodochrones of the travel times along the seismic line. For a color version of this figure, see www.iste.co.uk/cattin/himalaya1.zip

A regional trend of -1.58 $\mu$Gal/m obtained by (Hammer et al. 2013) is also taken into account. The final dataset consists of 139 corrected gravity measurements, which highlight variations along the profile (Figure 7.6). No change at the fault trace is observed, but a transition occurs at around 27 m north of the fault. The southern part of the profile is characterized by a moderate northward increase of approximately 4 $\mu$Gal/m. The northern part shows an increase twice as large with a northward increase of approximately 450 $\mu$Gal at 65 m.

## 7.2.2. *Case study data and inversion technique*

Geophysical inversion is a mathematical and statistical tool used to recover physical properties models from field observations and information on geological structures (Tarantola 2005). Inversion methods can be divided into deterministic and stochastic approaches. Deterministic inversion is a conventional linear approach that consists of gradually updating the model parameters to minimize the differences between observed and theoretical data computed inside the output model (Ellis and Oldenburg 1994). It is a model-driven inversion approach, relatively easy to implement. Still, strong a priori constraints are required to converge towards the best acceptable model. Unlike the deterministic approach (Ramirez et al. 2005), the stochastic inversion is a statistical process in which prior information and forward modeling are combined to produce different output models consistent with the available data. This approach provides a more complete description of the possible acceptable solutions but can be time-consuming. In this framework, Drukpa et al. (2017) propose a novel common stochastic approach to invert the near-surface geophysical data. Following Mosegaard and Tarantola (1995), a Markov Chain Monte Carlo technique is used to pseudo-randomly generate a large collection of models according to the posterior probability distribution.

This approach was applied to shallow geophysical data for the case study site in South Bhutan. Assuming a simplified geometry (Figure 7.7), each model is associated with only five bodies, including a south (STL) and north (NTL) top layers, a south (SL) and north (NL) shallow layers and a fault layer. Based on this formulation, a set of models results for a given dataset combines 10 parameter estimations that include either the velocity, the resistivity or the density of each body, as well as the thickness of layers, the fault location and the fault dip angle. In order to fix the limit of the solution-space, a priori parameters ranges for resistivity and velocity values are obtained from preliminary deterministic inversion using RES2DINV (Loke and Barker 1996) and RAYFRACT (Schuster 1993; Sheehan et al. 2005; Pasquet et al. 2015), respectively. The prior density contrasts between NL and the other bodies in the -500 to 500 kg/m$^3$ is assumed, while the prior information comes from structural and geomorphological observations (Long et al. 2011; Le Roux-Mallouf et al. 2016), which yield top layers thickness less than 5 m and a fault dip angle between 10° and 80°.

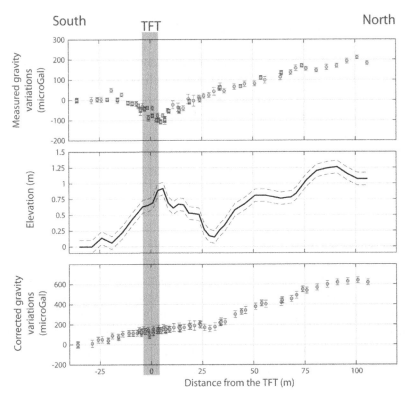

**Figure 7.6.** *Measured gravity (top), elevation (middle) and gravity variations corrected for both topographic effect and regional trend (bottom) along the study profile. Data uncertainties associated with both accuracy of the CG5 gravimeter and error in elevation measurement. For a color version of this figure, see www.iste.co.uk/ cattin/himalaya1.zip*

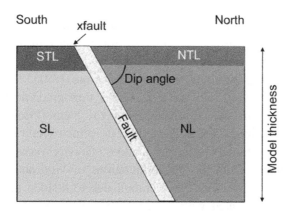

**Figure 7.7.** *Geometry of the model used in the stochastic inversion. STL – South Top layer, NTL – North Top Layer, SL – South Layer and NL – North Layer. xfault is the prior location of the fault as observed in the field. Model thickness is associated with the thickness investigated by each geophysical method. For a color version of this figure, see www.iste.co.uk/cattin/himalaya1.zip*

The pseudo-random walk through this multi-dimensional parameters space is controlled by the following rules for the transition between model $m_i$ to model $m_j$:

1) If $L(m_j) \geq L(m_i)$, then accept the proposed transition from $i$ to $j$.

2) If $L(m_j) < L(m_i)$, then accept the proposed transition from $i$ to $j$ with the probability $\frac{L(m_j)}{L(m_i)}$.

where $L(m_i)$ and $L(m_j)$ are the likelihood of the old and the new model, respectively. Here, we assume that the likelihood function can be written as:

$$L(m_i) = \exp\left(-\frac{1}{n_{obs}} \sum_{n=1}^{n_{obs}} \frac{|calc_n(m_i) - obs_n|}{\sigma_n}\right), \qquad [7.1]$$

where $n_{obs}$ is the number of data points, $obs$ is the data vector, and $\sigma$ is the total variance, i.e. the uncertainties associated with each data point. $calc(m_i)$ is the forward modeling function associated with the model $m_i$. This function is obtained using the different forward modeling depending on the considered datasets described below.

Two-dimensional geoelectrical modeling is performed with the software package R2 (Binley and Kemna 2005; Binley 2015). The current flow between electrodes is obtained using a quadrilateral mesh with an exponentially increasing node at depth and a constant node spacing in the horizontal direction. For seismic refraction, synthetic travel times are computed using the real receiver-shot configuration and solving the Eikonal equation with a finite-difference algorithm (Podvin and Lecomte 1991). Rays are traced in the obtained time field with the a posteriori time-gradient method. More precise travel times are then estimated along ray paths (Priolo et al. 2012). The model is discretized on a regular grid. The velocity field is parametrized by trilinear interpolation between grid nodes. Gravity variations along the profile are calculated from the 2D formulations of Won and Bevis (1987), which provide the gravitational acceleration due to n-sided polygons. Here, the polygons are associated with the geometry of the five bodies described above. The model is extended southward and northward to avoid edge effects at the two terminations.

The posterior probability of each model parameter is then obtained from the final collection of the $5 \times 10^5$ sampled models. Compared to the commonly deterministic approach, which leads to the more acceptable model, the main advantages of stochastic inversion include its ability (1) to assess the fault geometry because no smoothing is applied, (2) to provide a measurement of the uncertainties on the obtained dip angle and (3) to allow the study of trade-off analysis between geometric and either electrical resistivity, velocity or density properties. Using parallelism, the computation time associated with electrical, seismic and gravity inversion on a 10-core workstation is between approximately 10 and 24.5 hours.

## 7.3. Geophysical results of case study from south Bhutan

### 7.3.1. *Electrical resistivity tomography*

All dipole–dipole and Wenner–Schlumberger data were separately inverted using the stochastic approach. The set of most likely models derived from the stochastic approach explains the main features of the observed apparent resistivity pattern, except southward where some residual differences persist (Figure 7.8). It points out a high fault dip angle of approximately 70° (Figure 7.8). Bivariate frequency histograms indicate no tradeoff between dip angle and the other geometric and electrical parameters. This figure constrains the model parameters and discusses the robustness and relevance of the results.

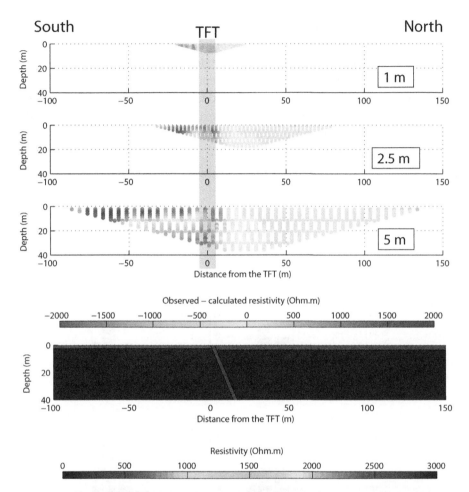

**Figure 7.8.** *Misfit between observed and calculated ERT pseudo-sections for electrode spacing of 1 m, 2.5 m and 5 m using WS configuration. This misfit is defined as the difference between the observed and calculated resistivities using the electrical model plotted at the bottom. For a color version of this figure, see www.iste.co.uk/ cattin/himalaya1.zip*

In southern Bhutan, the histograms suggest a 2.5-m thick fault zone. However, the resistivity of this unit remains poorly resolved. The inversion approach images thin low-resistive top layers, both on the southern (~2.5 m, ~550 $\Omega \cdot$m) and northern sides (~3.5 m, ~350 $\Omega \cdot$m). The small resistivity contrasts between those two top layers can prevent the estimation of the fault geometry at a very shallow depth (< 5 m). On the contrary, due to the very high resistivity contrast between the two deeper bodies (SL~3300 $\Omega \cdot$m vs.

NL~30 Ω·m), we consider the obtained fault dip angle as a well-constrained parameter down to 40 m depth. This assumption is confirmed by the narrow posterior distribution obtained for dip angle. Finally, some discrepancies between the observed and calculated pseudo-sections on the south part. Based on the simplified geometry of the model assuming horizontal layering, the stochastic inversion procedure cannot explain the north–south resistivity variations in the footwall of the TFT.

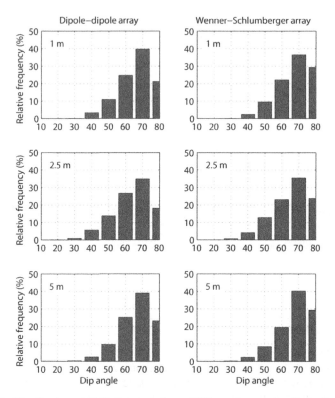

**Figure 7.9.** *Distribution of TFT dip angle from ERT sections using both dipole–dipole and Wenner–Schlumberger arrays. Electrode spacing ranges from 1 m (top) to 5 m (bottom). For a color version of this figure, see www.iste.co.uk/cattin/himalaya1.zip*

Altogether, this information on both geometry and resistivity contrast suggests an apparent resistivity contrast between both sides of the fault and a constant dip angle of ∼ 70° over a depth ranging between 5-40 m (Figure 7.9).

## 7.3.2. *Seismic tomography*

The set of final velocity models resulting from the stochastic inversion approach provides low travel-time residuals of ±3 ms in average for most of the source–receiver pairs (Figure 7.10). This suggests that assuming simple geometry captures most of the main features of the velocity field. Furthermore, travel time residuals show a relatively homogeneous pattern, except close to the fault trace between -5 and 15 m, where residuals abruptly increase from -5 ms to 4 ms northward. This result demonstrates that the presence of the fault influences seismic data.

Ray coverage (Figure 7.5) indicates a shallower resolution depth compared to ERT investigations. Resolution depth varies between the two sides of the fault, from approximately 8 m to approximately 5 m in the south and north, respectively. At these depths, the velocity models resulting from both the stochastic inversion (Figure 7.10) and tomography (Figure 7.5) point out high-velocity variations of about 50% at the transition of the fault zone. The 2D seismic model also emphasizes strong vertical velocity changes on both sides of the fault. High-velocity contrasts between top and bottom layers induce a concentration of rays at a depth between 2 and 4 m, which prevents deeper investigations, in particular in the north.

Taking into account this shallow investigation depth, the velocity field can be characterized by two deeper units of Vp~1,100 m/s (SL) and Vp~2,100 m/s (NL) below by two superficial low-velocity layers (STL: ~5 m, ~900 m/s and NTL: ~3 m, ~1,600 m/s). The stochastic inversion procedure also reveals that seismic data are sensitive to the dip angle parameter (Figure 7.11). The 2D seismic inversion result suggests a northward dipping fault with a very low-angle of approximately 20°-30° at depths down to approximately 5 m, which is consistent with field observations in the trench (Le Roux-Mallouf et al. 2020).

This observation is also in agreement with both ERT profile (Figure 7.3) and the related stochastic ERT results, which displays a change in dip angle with a more gentle slope of the TFT fault near the surface. Without information on the near-surface geometry of the fault, a constant fault dip angle was assumed, and the same model was used for the different ERT configurations.

Thereby the inversion procedure could not image this dip angle change near the surface with resistivity data only.

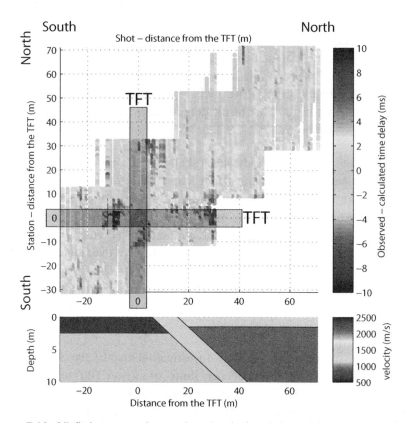

**Figure 7.10.** *Misfit between observed and calculated time delay corresponding to the travel time residuals. This misfit is defined as the difference between the observed and calculated travel time using the velocity model plotted at the bottom. Compared to resistivity model (Figure 7.8), we can noted the limit of the y-axis, which corresponds to a lower depth of investigation. For a color version of this figure, see www.iste.co.uk/cattin/himalaya1.zip*

Hence, seismic data inversion with field observations confirms a northward dipping fault with a low angle of approximately 20°–30° at very shallow depth. Seismic and electrical resistivity images together suggest a dip angle that increases gradually up to approximately 70° at a depth of 5–10 m.

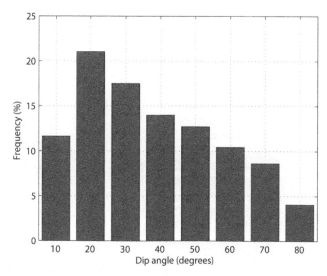

**Figure 7.11.** *Distribution of TFT dip angle obtained from seismic measurements. For a color version of this figure, see www.iste.co.uk/cattin/himalaya1.zip*

### 7.3.3. *Micro-gravity*

Since gravity measurements are affected mainly by the deeper part of the model, the focus of the study is on the long wavelength of the gravity profile associated with the two south–north gravity gradients (Figure 7.6). The result of the stochastic inversion suggests that the observed northward increase of gravity measurements is mostly related to both $\Delta\varrho$ the density contrast between SL and NL and $\alpha$ the fault dip angle (Figure 7.12). As indicated in Figure 7.12, gravity measurements cannot be used to assess the other density and geometric parameters, which remain poorly constrained.

The gravity model result reveals a tradeoff between $\Delta\varrho$ and $\alpha$: the higher the density contrast, the lower the fault dip angle. For $\Delta\varrho$= -350 kg/m$^3$, the fault dip angle is approximately 30° , whereas for $\Delta\varrho$= -200 kg/m$^3$, the fault dip angle is approximately 60° (Figures 7.12a and b). This leads to a wide distribution of the fault dip angle (Figure 7.12c). The maximum obtained at $\alpha \sim$ 30–40° and the most likely model thickness down to an 80 m depth suggests a fault that flattens at depths below 40 m.

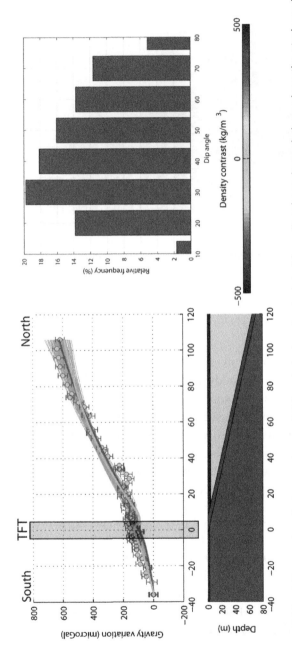

**Figure 7.12.** a) Comparison between observed (blue circles) and calculated (gray lines) gravity variations along the study profile obtained for the 100 best-fitting models. (b) Density contrast models associated with the red (30° dip angle) and green lines (60° dip angle) plotted above. (c) Distribution of TFT dip angle obtained from gravity stochastic inversion. The red dots indicate the model with the lower misfit value. For a color version of this figure, see www.iste.co.uk/cattin/himalaya1.zip

## 7.4. Implications of near-surface geophysical findings

The geometry of the fault, especially at shallow depth, is a crucial parameter for better understanding deformation kinematics and accommodation at crustal scale. In particular, the slip rate can be estimated by combining the subsurface dip angle and terrace dating results. The new constraints for the TFT geometry deduced from near-surface geophysical techniques allow studying stress partitioning at the frontal thrust zone and its associated seismic hazard implications in south Central Bhutan.

### 7.4.1. *Subsurface imaging*

Taking advantage of the various scales of investigation coming from ERT, seismic and gravity methods, an accurate description of shallow structures and fault geometry at depth is obtained in the case study. The subsurface can be subdivided into three main zones:

1) a very shallow part up to 5 m depth well constrained by both field observations and seismic data considering the ray coverage;

2) an intermediate depth part well-imaged by ERT sections between 5 and 40 m depth due to high resistivity contrasts;

3) a deeper part documented by gravity measurements below 40 m depth.

The fault geometry discussed here arises from the integration of these three surface sensitivities.

In terms of lithological setting and water content, the geophysical datasets suggest a thin layer ($\sim$3-5 m) that appears to be present on both sides of the fault trace and probably corresponds to recent alluvial deposits. Along the profile, resistivity and velocity variations at shallow depth may be due to a northward decrease of water saturation. Below these superficial layers, in the hanging wall of the TFT, the obtained very low-resistivity values ($< 30\ \Omega \cdot$m), the high $V_p$ of approximately 2,100 m/s and the relatively low densities probably underline a phyllite unit, which can be observed in the field.

Overall, the geophysical methods image a more complex fault geometry than proposed by earlier studies (Berthet et al. 2014; Le Roux-Mallouf et al. 2016). The geophysical results show a TFT with a flat and listric-ramp geometry with a low dip angle of 20° –30° at shallow depth, steeply dipping at ∼70° in the middle and gradually flattening to a shallower dip angle of 30° –40° in its deeper part (Figure 7.13).

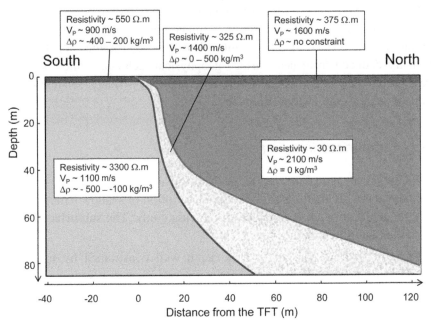

**Figure 7.13.** *Simplified cross-section showing the main geophysical results obtained from electrical resistivity tomography, seismic refraction and gravity measurements. Together these results suggest a TFT with a flat and ramp geometry, with a surface dip angle of ∼ 20° reaching ∼ 70° at 20 m depth and flattening in its deeper part. Note that dashed area is bounded by the two end-member models of fault geometry given by the green and the red lines. Hence, this area does not represent the fault thickness, which is estimated to 2.5 m. For a color version of this figure, see www.iste.co.uk/cattin/himalaya1.zip*

### 7.4.2. *Overthrusting slip rate assessment*

Berthet et al. (2014) estimated a Holocene vertical slip rate of 8.8±2.1 mm/yr by dating two uplifted river terraces in Sarpang area. Assuming a dip angle of 20–30°, a slip rate of 20.8±8.8 mm/yr is estimated, which is consistent with the GPS convergence rate of 17 mm/yr obtained across central Bhutan (Marechal et al. 2016). Finally, they conclude that the TFT mainly accommodates the Himalayan convergence. However, this major conclusion can be revisited in light of our new constraints on the TFT geometry.

First, assuming a constant overthrusting slip rate along the TFT, a vertical velocity profile is calculated from this observed uplift rate (Okada 1985). As expected, this calculated profile depends on TFT geometry: a higher fault dip angle implies a higher uplift rate. More surprisingly, it also depends on the distance between the TFT and the location of dated samples. For instance, a distance of 5 m from the TFT yields two very different vertical velocity profiles associated with the two end-member models for the fault geometry. On the contrary, if the uplift rate is measured about 10 m north from the TFT, the uplift rate difference drastically reduces. In other words, due to the flat and listric-ramp geometry of the shallow TFT, the uplift rate measured on the top of river terraces is spatially variable and cannot be constant. This result questions the validity of commonly used approaches for which a mean uplift rate is obtained by combining several uplifted terraces located at various distances from the front. Furthermore, assuming that the far-field GPS shortening rate corresponds to an upper limit for the uplift rate, our calculation shows that part of the models are unrealistic. This suggests that both the convergence rates derived from GPS and the uplift measurements can be used to reduce the a priori geometric parameter ranges tested in our stochastic approach.

Second, assuming no prior information on the relative location of uplift rate measurements, the overthrusting slip rate can be deduced from the TFT geometry. The slip rate associated with a rigid block model with a constant dip angle $\alpha$ can be easily estimated from

$$\text{slip rate} = \frac{\text{uplift rate}}{\sin(\alpha)} \qquad [7.2]$$

As previously proposed by Berthet et al. (2014), this simple approach gives a minimum dip angle of 30° for which most of the convergence across central Bhutan is accommodated along the TFT. However, the steeper is the dip angle; the greater is the chance for slip partitioning with other faults. Assuming a constant uplift rate of 8.8±2.1 mm/yr associated with no information on the sampling location, the slip rate can be estimated from less straightforward modeling based on the obtained geometry. In that case, using dislocations embedded in a homogeneous half-space (Okada 1985), the obtained slip rate exhibits high variations along the profile from 20 to 40 mm/yr above the very shallow part of the fault to 10 to 20 mm/yr in the northern part of the profile. Using the convergence rate as a maximum value for the slip, this result suggests a minimum distance of 8 m for the steepening of the TFT and an accommodation of at least $10\pm2$ mm/yr of the 17 mm/yr of convergence at the TFT.

The obtained uncertainties associated with this slip rate estimate arise mainly from both the location of samples for terrace dating and the fault geometry inferred from geophysical inversion.

### 7.4.3. *Deformation at the topographic front*

Based on new constraints on the TFT geometry and the resulting slip rate, it is proposed that at least 60% of the convergence rate due to ongoing underthrusting of India beneath the Himalaya is accommodated by the TFT.

It results that additional faults must be active in this area, which is consistent with results obtained by Dey et al. (2016) in the Kangra section of the Indian Himalaya where, besides the MFT, other out-of-sequence faulting such as the Jwalamukhi Thrust (JMT) accommodates part of the Sub-Himalayan shortening. In the case study area in south Bhutan, we can mention either the north-propagating emerging thrust front (FBT) documented by Dasgupta et al. (2013) in the Brahmaputra plain or the Main Boundary Thrust (MBT), which accommodates the present-day deformation in eastern Bhutan (Marechal et al. 2016).

Based on the recent studies, it is now well established that at least two major earthquakes have occurred on the TFT in the past, the final one having occurred about 300 years ago in 1714 (Hetényi et al. (2016), Le Roux-Mallouf et al. (2016), see Volume 3 – Chapter 5). Thus, a slip deficit of 3–5 m has accumulated on the TFT during this interseismic period, and could potentially be released in a large magnitude earthquake with high probability of rupture reaching the surface (see Volume 3 – Chapter 5).

## 7.5. Conclusion

This chapter presents high-resolution near-surface geophysical imaging results based on a joint approach, including electrical resistivity, seismic and gravity data to constrain the TFT geometry in south Central Bhutan.

Our results show that a flat and listric-ramp geometry characterizes the upper part of the TFT with high variations of dip angle. This geometry differs from the constant fault dip angle inferred from surface observation only. Estimating the slip rate without additional depth constraints can induce significant errors, arising from the terrace dating to determine the uplift rate and the projection of the fault dip angle based on surface observations.

By combining information from surface observations with our new constraints on the fault geometry, we estimate that at least 60% of the Himalayan convergence is accommodated by the TFT, making this fault a high seismic hazard zone.

The hypothesis of slip partitioning cannot be totally ruled out. Other faults such as the FBT emerging in the Brahmaputra plain, and the MBT can also be active. Therefore, further studies combining geomorphology and near-surface geophysics along the Himalayan front, especially towards the eastern part of Bhutan, will be helpful to study potential lateral variations in the fault geometry and its implication on the present-day strain partitioning. Moreover, local variability across the TFT may be assessed by exploring areas within a few 100s of meters along strike with respect to the Sarpang study area.

## 7.6. References

Berthet, T., Ritz, J.F., Ferry, M., Pelgay, P., Cattin, R., Drukpa, D., Braucher, R., Hetényi, G. (2014). Active tectonics of the eastern Himalaya: New constraints from the first tectonic geomorphology study in southern Bhutan. *Geology*, 42(5), 427–430.

Binley, A. (2015). *Tools and Techniques: Electrical Methods*. Elsevier, Amsterdam.

Binley, A. and Kemna, A. (2005). Electrical methods. In *Hydrogeophysics*, Hubbard, S. and Rubin,Y. (eds). Springer, Dordrecht.

Boness, N.L. and Zoback, M.D. (2004). Stress-induced seismic velocity anisotropy and physical properties in the SAFOD Pilot Hole in Parkfield, CA. *Geophysical Research Letters*, 31, L15S17, doi:10.1029/2003GL019020.

Cattin, R., Mazzotti, S., Baratin, L.M. (2015). GravProcess: An easy-to-use MATLAB software to process campaign gravity data and evaluate the associated uncertainties. *Computers and Geosciences*, 81, 20–27.

Dahlin, T. and Zhou, B. (2004). A numerical comparison of 2D resistivity imaging with 10 electrode arrays. *Geophysical Prospecting*, 52(5), 379–398.

Dasgupta, S., Mazumdar, K., Moirangcha, L.H., Gupta, T.D., Mukhopadhyay, B. (2013). Seismic landscape from Sarpang re-entrant, Bhutan Himalaya foredeep, Assam, India: Constraints from geomorphology and geology. *Tectonophysics*, 592, 130–140.

Demanet, D., Pirard, E., Renardy, F., Jongmans, D. (2001). Application and processing of geophysical images for mapping faults. *Computers and Geosciences*, 27(9), 1031–1037.

Dey, S., Thiede, R.C., Schildgen, T.F., Wittmann, H., Bookhagen, B., Scherler, D., Strecker, M.R. (2016). Holocene internal shortening within the northwest Sub-Himalaya: Out-of-sequence faulting of the Jwalamukhi Thrust, India. *Tectonics*, 35(11), 2677–2697.

Drukpa, D., Stephanie, G., Rodolphe, C., Kinley, N., Nicolas, L.M. (2017). Impact of near-surface fault geometry on secular slip rate assessment derived from uplifted river terraces: Implications for convergence accommodation across the frontal thrust in southern Central Bhutan. *Geophysical Journal International*, 212(2), 1315–1330.

Ellis, R.G. and Oldenburg, D.W. (1994). Applied geophysical inversion. *Geophysical Journal International*, 116(1), 5–11.

Gelis, C., Revil, A., Cushing, M., Jougnot, D., Lemeille, F., Cabrera, J., de Hoyos, A., Rocher, M. (2010). Potential of electrical resistivity tomography to detect fault zones in limestone and argillaceous formations in the experimental platform of tournemire, France. *Pure and Applied Geophysics*, 167, 1405–1418.

Hammer, P., Berthet, T., Hetényi, G., Cattin, R., Drukpa, D., Chophel, J., Lechmann, S., Moigne, N.L., Champollion, C., Doerflinger, E. (2013). Flexure of the India plate underneath the Bhutan Himalaya. *Geophysical Research Letters*, 40(16), 4225–4230.

Hetényi, G., Cattin, R., Berthet, T., Le Moigne, N., Chophel, J., Lechmann, S., Hammer, P., Drukpa, D., Sapkota, S.N., Gautier, S. et al. (2016). Segmentation of the Himalayas as revealed by arc-parallel gravity anomalies. *Scientific Reports*, 6(September), 33866.

Hung, J.-H., Ma, K.-F., Wang, C.-Y., Ito, H., Lin, W., Yeh, E.-C. (2009). Subsurface structure, physical properties, fault-zone characteristics and stress state in scientific drill holes of Taiwan Chelungpu Fault Drilling Project. *Tectonophysics*, 466(3), 307–321.

Jeppson, T.N., Bradbury, K.K., Evans, J.P. (2010). Geophysical properties within the San Andreas Fault Zone at the San Andreas Fault Observatory at Depth and their relationships to rock properties and fault zone structure. *Journal of Geophysical Research: Solid Earth*, 115, B12423, doi:10.1029/2010JB007563.

Kaiser, A.E., Green, A.G., Campbell, F.M., Horstmeyer, H., Manukyan, E., Langridge, R.M., McClymont, A.F., Mancktelow, N., Finnemore, M., Nobes, D.C. (2009). Ultrahigh-resolution seismic reflection imaging of the Alpine Fault, New Zealand. *Journal of Geophysical Research: Solid Earth*, 114(11), 1–15.

Le Roux-Mallouf, R., Ferry, M., Ritz, J.-F., Berthet, T., Cattin, R., Drukpa, D. (2016). First paleoseismic evidence for great surface-rupturing earthquakes in the Bhutan Himalayas. *Journal of Geophysical Research: Solid Earth*, 121(10), 7271–7283.

Le Roux-Mallouf, R., Ferry, M., Cattin, R., Ritz, J., Drukpa, D. (2020). A 2600-yr-long paleoseismic record for the Himalayan Main Frontal Thrust (Western Bhutan). *Solid Earth*, 11, 2359–2375

Loke, M. (2015). Tutorial: 2-D and 3-D electrical imaging surveys. *Geotomo Software Malaysia*, July, 176.

Loke, M. and Barker, R.D. (1996). Rapid least-squared inversion of apparent resisitivity pseudosections by a quasi-Newton method. *Geophysical Prospecting*, 44(1), 131–152.

Long, S., McQuarrie, N., Tobgay, T., Grujic, D., Hollister, L. (2011). Geologic map of Bhutan. *Journal of Maps*, 7(1), 184–192.

Louis, I.F., Raftopoulos, D., Goulis, I., Louis, F.I. (2002). Geophysical imaging of faults and fault zones in the urban complex of Ano Liosia neogene basin, Greece: Synthetic simulation approach and field investigation. *International Conference on Earth Sciences and Electronics*, October, 269–285.

Marechal, A., Mazzotti, S., Cattin, R., Cazes, G., Vernant, P., Drukpa, D., Thinley, K., Tarayoun, A., Roux-Mallouf, L., Thapa, B.B. et al. (2016). Evidence of interseismic coupling variations along the Bhutan Himalayan arc from new GPS data. *Geophysical Research Letters*, 43. doi:10.1002/2016GL071163.

Morandi, S. and Ceragioli, E. (2002). Integrated interpretation of seismic and resistivity images across the "Val d'Agri" graben (Italy). *Annals of Geophysics*, 45(2), 259–271.

Mosegaard, K. and Tarantola, A. (1995). Monte Carlo sampling of solutions to inverse problems. *Journal of Geophysical Research: Solid Earth*, 100(B7), 12431–12447.

Mussett, A., Khan, M., Button, S. (2000). *Looking Into the Earth: An Introduction to Geological Geophysics*. Cambridge University Press.

Nguyen, F. (2005). Near-surface geophysical imaging and detection of slow active faults. Doctorate, Department of Georessources, Geotechnologies and Construction Materials, May, 359.

Nguyen, F., Garambois, S., Jongmans, D., Pirard, E., Loke, M.H. (2005). Image processing of 2D resistivity data for imaging faults. *Journal of Applied Geophysics*, 57(4), 260–277.

Nguyen, F., Garambois, S., Chardon, D., Hermitte, D., Bellier, O., Jongmans, D. (2007). Subsurface electrical imaging of anisotropic formations affected by a slow active reverse fault, Provence, France. *Journal of Applied Geophysics*, 62(4), 338–353.

Okada, Y. (1985). Surface deformation due to shear and tensile faults in a half-space. *Bulletin of the Seismological Society of America*, 75(4), 1135–1154.

Pasquet, S., Bodet, L., Dhemaied, A., Mouhri, A., Vitale, Q., Rejiba, F., Flipo, N., Guérin, R. (2015). Detecting different water table levels in a shallow aquifer with combined p-, surface and sh-wave surveys: Insights from vp/vs or poisson's ratios. *Journal of Applied Geophysics*, 113, 38–50.

Phillips, W.J. and Kuckes, A.F. (1983). Electrical conductivity structure of the San Andreas fault in central California. *Journal of Geophysical Research: Solid Earth*, 88(B9), 7467–7474.

Podvin, P. and Lecomte, I. (1991). Finite difference computation of traveltimes in very contrasted velocity models: A massively parallel approach and its associated tools. *Geophysical Journal International*, 105(1), 271–284.

Priolo, E., Lovisa, L., Zollo, A., Böhm, G., D'Auria, L., Gautier, S., Gentile, F., Klin, P., Latorre, D., Michelini, A. et al. (2012). The Campi Flegrei blind test: Evaluating the imaging capability of local earthquake tomography in a volcanic area. *International Journal of Geophysics*, 505286, 37, https://doi.org/10.1155/2012/505286.

Ramirez, A.L., Nitao, J.J., Hanley, W.G., Aines, R., Glaser, R.E., Sengupta, S.K., Dyer, K.M., Hickling, T.L., Daily, W.D. (2005). Stochastic inversion of electrical resistivity changes using a Markov Chain Monte Carlo approach. *Journal of Geophysical Research: Solid Earth*, 110, B02101. doi:10.1029/2004JB003449.

Reynolds, J. (1997). *An Introduction to Applied and Environmental Geophysics*. John Wiley & Sons, Hoboken, NJ.

Schuster, G.T. (1993). Wavepath eikonal traveltime inversion: Theory. *Geophysics*, 58(9), 1314.

Sheehan, J.R., Doll, W.E., Mandell, W.A. (2005). An evaluation of methods and available software for seismic refraction tomography analysis. *Journal of Environmental & Engineering Geophysics*, 10(1), 21–34.

Suzuki, K., Toda, S., Kusunoki, K., Fujimitsu, Y., Mogi, T., Jomori, A. (2000). Case studies of electrical and electromagnetic methods applied to mapping active faults beneath the thick quaternary. *Engineering Geology*, 56(1), 29–45.

Tarantola, A. (2005). *Inverse Problem Theory and Methods for Model Parameter Estimation*. Society for Industrial and Applied Mathematics, Philadelphia, PA.

Telford, W.M., Geldart, L.P., Sheriff, R.E. (1990). *Applied Geophysics, Vol. 1*. Cambridge University Press.

Villani, F., Tulliani, V., Sapia, V., Fierro, E., Civico, R., Pantosti, D. (2015). Shallow subsurface imaging of the Piano di Pezza active normal fault (central Italy) by high-resolution refraction and electrical resistivity tomography coupled with time-domain electromagnetic data. *Geophysical Journal International*, 203(3), 1482–1494.

Wise, D.J., Cassidy, J., Locke, C.A. (2003). Geophysical imaging of the Quaternary Wairoa North Fault, New Zealand: A case study. *Journal of Applied Geophysics*, 53(1), 1–16.

Won, I.J. and Bevis, M. (1987). Computing the gravitational and magnetic anomalies due to a polygon: Algorithms and Fortran subroutines. *Geophysics*, 52(2), 232–238.

# 8

# Overview of Hydrothermal Systems in the Nepal Himalaya

**Frédéric GIRAULT[1], Christian FRANCE-LANORD[2],
Lok Bijaya ADHIKARI[3], Bishal Nath UPRETI[4],
Kabi Raj PAUDYAL[5], Ananta Prasad GAJUREL[5],
Pierre AGRINIER[1], Rémi LOSNO[1], Sandeep THAPA[1],
Shashi TAMANG[6,1,5], Sudhan Singh MAHAT[7],
Mukunda BHATTARAI[1], Bharat Prasad KOIRALA[3],
Ratna Mani GUPTA[3,8], Kapil MAHARJAN[3],
Nabin Ghising TAMANG[3], Hélène BOUQUEREL[1],
Jérôme GAILLARDET[1], Mathieu DELLINGER[1],
François PREVOT[1], Carine CHADUTEAU[1], Thomas RIGAUDIER[2],
Nelly ASSAYAG[1] and Frédéric PERRIER[1]**

[1] *Paris Cité University, France*
[2] *University of Nancy, Vandoeuvre-lès-Nancy, France*
[3] *Department of Mines and Geology, Kathmandu, Nepal*
[4] *Nepal Academy of Science and Technology, Kathmandu, Nepal*
[5] *Tribhuvan University, Kathmandu, Nepal*
[6] *University of Turin, Italy*
[7] *Sanjen Jalavidhyut Company Limited, Kathmandu, Nepal*
[8] *Academia Sinica, Taipei, Taiwan*

## 8.1. Introduction

Convergent plate boundaries are essential components of the global carbon dioxide ($CO_2$) balance of the Earth (Kerrick and Caldeira 1998). Their role

*Himalaya, Dynamics of a Giant 1,*
coordinated by Rodolphe CATTIN and Jean-Luc EPARD.
© ISTE Ltd 2023.

of atmospheric $CO_2$ sink (Mörner and Etiope 2002) is controlled by silicates weathering (Gaillardet et al. 1999; Wolff-Boenisch et al. 2009). However, large active orogens also produce and release $CO_2$-rich fluids (Irwin and Barnes 1980), through decarbonation of carbonates (Ague 2000; Evans et al. 2004) and sulfide oxidation (Torres et al. 2014; Kemeny et al. 2021). Although the ratio emission/sink of carbon is believed to be relatively close to 1, it is difficult to argue whether a steady state is reached or not; while the sink term appears relatively stable (Märki et al. 2021), some sources are still unconstrained. Indeed, although the emissions of $CO_2$ associated with active volcanism are well studied (e.g. Burton et al. 2013), there is a lack of observations and measurements of carbon emissions associated with the formation of mountain belts (Gaillardet and Galy 2008), hence hampering proper evaluation of the global carbon budget. Major orogenic thrust fault zones dynamically interrelate fluid circulation and Earth's crust deformation and earthquakes potentially through crustal permeability (Manning and Ingebritsen 1999; Ingebritsen and Manning 2010; Manga et al. 2012; Miller 2020). Orogenic $CO_2$ may indeed reveal sensitive to the long-term strain mountain buildup and the short-term transient stress release during earthquakes, of potential interest in such highly populated seismogenic areas on Earth (Adhikari 2021).

The Nepal Himalaya appears as a natural laboratory for the study of current spatial and temporal variations of $CO_2$ release at large scale. Indeed, first, high seismicity is concentrated on a mid-crustal décollement ramp (Pandey et al. 1995, 1999; Avouac et al. 2001, and Volume 1 – Chapter 4) below the Main Central Thrust (MCT) shear zone (Upreti 1999), the Main Himalayan Thrust (Nábělek et al. 2009), that accommodates half of the convergence ($2 \text{ cm.year}^{-1}$) between Indian Plate and Tibet (Avouac 2003; Ader and Avouac 2012). Second, in the MCT zone, fluid occurrence at crustal depth could be suspected by the high electrical conductivity body inferred from magnetotelluric soundings (Lemonnier et al. 1999; Unsworth et al. 2005). Third, the monsoon regime impacts deformation and seismicity, with higher surface load in summer (Bettinelli et al. 2008; Chanard et al. 2014) and enhanced seismicity in winter (Bollinger et al. 2004), as well as the activity of seismic swarms (Adhikari et al. 2021) and possibly the timing of large earthquakes (Panda et al. 2018), suggesting interrelated relationships with crustal fluids and surface hydrological forcing. The MCT zone in the Nepal Himalaya exhibits a large number of hydrothermal systems, where various surface manifestations such as thermal springs, travertine deposits, hydrothermal alteration, "tectonic" fumaroles and diffuse degassing structures can be studied. In such a highly segmented seismogenic area prone to mega-earthquakes (e.g. Bilham 2019), it appears particularly important to investigate the along-strike heterogeneity of current $CO_2$ emissions and thermal spring characteristics, and to monitor such geochemical parameters as a function of time to be able to test, ultimately, their potential association with crustal deformation and the Himalayan seismic cycle (see Volume 3 – Chapters 5 and 6).

Several thermal springs, famous pilgrimage sites for the local population, were first reported in Central and Western Nepal in the vicinity of the MCT and Main Frontal Thrust (MFT) (Le Fort 1975; Bhattarai 1980; Grabczak and Kotarba 1985; Bhattarai 1986; Kotarba 1986; Sharma 1995; Perrier et al. 2002a). Then, high alkalinity of thermal springs was shown to contribute in a large part to the total dissolved inorganic carbon (DIC) of the entire Narayani watershed (Evans et al. 2004). The high carbon isotopic compositions ($\delta^{13}C$) of the thermal springs suggested a metamorphic decarbonation source and a massive $CO_2$ degassing near the water table in the Trisuli and Marsyandi valleys, Central Nepal (Becker 2005; Becker et al. 2008; Evans et al. 2008). Concomitantly, a high thermal gradient was inferred from water chemistry in the Trisuli valley (Derry et al. 2009). First direct evidence of gaseous $CO_2$ emissions on the ground was reported in the Trisuli valley, at the Syabru-Bensi hydrothermal system (Perrier et al. 2009). These seminal studies have stimulated active field studies to characterize thermal springs and quantify $CO_2$ emissions in the Trisuli valley (Girault et al. 2014a, 2018b) and in other north–south valleys in the MCT zone in Western, Mid-Western and Central Nepal (Girault et al. 2014b, 2018b; Ghezzi et al. 2017, 2019). This $CO_2$-rich gas is characterized by radiogenic helium, high radon-222 content (radioactive gas of 3.82 day half-life) and $\delta^{13}C$ values suggesting metamorphic $CO_2$ production at more than 5 km depth. A 110-km-long region with particularly large $CO_2$ emissions was identified between the Annapurnas and Langtang ranges, coincident with a seismic gap since the 1936 Rukum earthquake (Szeliga et al. 2010), suggesting possible link between crustal permeability and earthquakes (Girault et al. 2014b).

On April 25, 2015, this high $CO_2$-emitting segment ruptured with the deadly Mw 7.9 Gorkha earthquake (Adhikari et al. 2015; Galetzka et al. 2015).

This earthquake released only partially the stored elastic energy sufficient to produce a possible Mw $\sim$9 mega-earthquake (Stevens and Avouac, 2016; see Volume 3 – Chapter 8). Recently, hydrothermal unrest and significant $CO_2$ emission changes have been reported following the 2015 Gorkha earthquake in Central Nepal (Girault et al. 2018c), opening the way towards the use of these sensitive natural geosystems as sentinels of the state of the Himalayan seismic cycle.

In this chapter, we present an up-to-date overview of the known hydrothermal systems in the Nepal Himalaya. After a brief description of the measurement methods, we introduce the main hydrothermal sites and present the associated results separately for thermal springs and gaseous emission zones from Far-Western to Eastern Nepal in the vicinity of the MCT and Main Frontal Thrust (MFT).

## 8.2. Measurement methods

In this section, we briefly recall the different measurement techniques used to identify, detect and measure various parameters in thermal spring waters and gaseous emissions. Every uncertainty is given around one-sigma standard deviation (68% confidence level) and averages are weighted arithmetic means except stated otherwise.

### 8.2.1. *Exploration approach*

Before field investigations, it appeared always essential to document the site and, when possible, contact local people. Topographic maps and sometimes touristic trekking maps depicted locations of the main hot springs with varying accuracy. However, the locations of secondary hot springs or even important hot springs in remote locations were not given. In addition, gaseous emissions were never indicated.

In general, it appeared easier to identify the presence of a hot spring, because they were known and sometimes used by the local inhabitants. In some cases, thermal infrared imaging or the presence of pervasive odor of hydrogen sulfide ($H_2S$) helped us to discover small hot water seepages.

The detection of gaseous $CO_2$ emissions was more difficult. We generally employed a combination of various tools and approaches, such as the use of thermal infrared imaging, the presence of one or several hot springs nearby, of a pervasive odor of $H_2S$, the absence of vegetation or the presence of specific species, the measurement of high radon-222 flux, surface temperature anomalies, the occurrence of water bubbles, the presence of cavities, the occurrence of inactive or active travertine deposits and discussions with local people.

In some cases, even after years of investigations in a given valley, it was possible to discover new springs and gaseous $CO_2$ emissions. This fact suggests that numerous hot springs or gaseous emission zones may still be unknown or unidentified and that our current approach leading to this crustal fluids overview is the first step towards a more exhaustive picture of thermal spring and gaseous emission occurrences in the Nepal Himalaya.

### 8.2.2. *Thermal spring water measurements*

Water temperature was measured directly at the mouth of each spring using digital thermometers (high-precision Pt Fisherbrand™ Traceable™ thermometer, USA; Generic TP101 Digital Thermometer, China) regularly inter-calibrated in the laboratory with a reference thermometer (Digital Thermometer model 4400 Ertco Eutechnics, USA) and inter-compared with high-precision ($10^{-3}$°C) and high-sensitivity ($10^{-4}$°C) thermometers (Seabird™ 39plus, Sea-Bird Scientific, USA). Experimental uncertainty of a given measurement was ±0.1°C.

Water pH was measured in the field using pH meters (H170 Portable pH meter, Hach, Germany; HI98107 and HI98130 pH meters, Hanna Instruments, USA), systematically recalibrated in the field using buffer solutions before any measurement. Experimental uncertainty of a given measurement was ±0.1.

Water electrical conductivity was measured in the field using conductivity meters (HI-98304 conductivity meter, Hanna Instruments, USA), systematically recalibrated in the field using buffer solutions before any measurement. Experimental uncertainty of a given measurement was ±20 $\mu$S.cm$^{-1}$.

Water flowrate was determined using a stopwatch and measuring cylinders or calibrated buckets and was repeated at least three times.

Dissolved radon-222 concentration in water was measured in the field by emanometry in air (Girault et al. 2018a). After reaching air–water equilibrium by manual shaking, radon concentration was inferred from scintillation flask sampling and photomultiplier counting (CALEN™, Algade, France), as described elsewhere (Girault et al. 2018b). Expressed in $Bq.L^{-1}$, experimental uncertainty ranged from 5 to 30%. Dissolved radium-226 concentration in water was similarly measured in the IPGP laboratory after keeping the bottle closed for at least 50–80 days (Perrier et al. 2016). Expressed in $mBq.L^{-1}$, experimental uncertainty was similar to that of dissolved radon concentration.

Major and trace elements in water were determined in the laboratory in France on 15 mL filtered (at 0.2 or 0.45 $\mu$m) and acidified (few droplets of nitric acid at 15.8 M) water samples collected in the field using spectrometers (ICP-AES and ICP-MS at IPGP, Université de Paris, CRPG and Université de Montpellier). Various international standards and blanks were used. Samples measured using different spectrometers gave similar results with differences better than 15% on average. Mean relative experimental uncertainties were 3% for major elements and 15% for trace elements.

### 8.2.3. *$CO_2$ flux and radon flux measurements*

Surface $CO_2$ flux was measured using the accumulation chamber method (Chiodini et al. 1998). The increase in $CO_2$ concentration in the chamber was measured using portable infrared $CO_2$ sensors (Testo™ 535, Testo AG, Germany; Airwatch™ PM 1500, Geotechnical Instruments Ltd., UK; Vaisala™ CARBOCAP® Hand-Held GM70, Finland), regularly inter-calibrated in the laboratory. Assessment over six orders of magnitude of $CO_2$ flux, expressed in $g.m^{-2}.d^{-1}$, is based on numerous systematic tests and our 15-year-long experience in various conditions including remote locations and monsoon periods (Girault et al. 2009, 2014a, 2018b). Relative experimental uncertainty varied on average from 10% for values smaller than 100 $g.m^{-2}.d^{-1}$ to 30% for values larger than 5,000 $g.m^{-2}.d^{-1}$. The total $CO_2$ discharge of a given site, expressed in $mol.s^{-1}$, was estimated using the $CO_2$ flux dataset by kriging and interpolation procedures (Girault et al. 2014a).

Surface radon-222 (hereinafter radon) flux was measured using the accumulation chamber method (Girault et al. 2009; Perrier et al. 2009). The increase in radon activity concentration in the chamber was measured after sampling using scintillation flasks (Algade, France) and counting 3.5 hours after sampling using photomultipliers (CALEN™, Algade, France), regularly inter-calibrated in the laboratory. Assessment over five orders of magnitude of radon flux, expressed in $10^{-3}$ $Bq.m^{-2}.s^{-1}$, is based on numerous systematic tests and our 15-year-long experience in various conditions including remote locations and monsoon periods (Richon et al. 2011; Girault et al. 2014a, 2018b). Relative experimental uncertainty varied on average from 15% for fluxes of $100\times10^{-3}$ $Bq.m^{-2}.s^{-1}$ to 30% for fluxes of $10\times10^{-3}$ and $10,000\times10^{-3}$ $Bq.m^{-2}.s^{-1}$. The total radon discharge of a given site, expressed in $Bq.s^{-1}$, was estimated using the radon flux dataset by kriging and interpolation procedures (Girault et al. 2014a).

### 8.2.4. *Carbon content and isotopic composition measurements*

Dissolved inorganic carbon (DIC) content and carbon isotopic composition were measured in the laboratory on water samples collected in the field in 12 mL glass screw-cap vials. DIC concentration ($C_{DIC} = [H_2CO_3] + [HCO_3^-] + [CO_3^{2-}]$) and isotopic composition ($\delta^{13}C_{DIC}$), expressed in mol $L^{-1}$ and in per mil relative to the standard values of Vienna Pee Dee Belemnite (V-PDB), respectively, were determined using a gas chromatograph coupled to an isotope ratio mass spectrometer (GCIRMS, GV 2003, GV Instruments, UK) at IPGP (Paris, France) (Assayag et al. 2006). Relative experimental uncertainty of $C_{DIC}$ was ±1–2%; experimental uncertainty of $\delta^{13}C_{DIC}$ was ±0.1 ‰.

Gaseous $CO_2$ content and carbon isotopic composition were measured on gas samples, collected in evacuated glass tubes in the field from surface $CO_2$ emission zones or from bubbles in water. After removing water and condensable gases, the $CO_2$ fraction of gas samples was determined manometrically with a mean experimental uncertainty of 0.1 %. The $\delta^{13}C$ of $CO_2$ of the gas samples ($\delta^{13}C_{gas}$ or $\delta^{13}C_{bubbles}$), expressed in per mil relative to the standard values of V-PDB, was measured after off-line purification by mass spectrometry on a Finnigan™ MAT-253 mass spectrometer (Thermo Electron Corp., Germany) at CRPG (Nancy, France) (Evans et al. 2008; Girault et al. 2014a). Experimental uncertainty was ±0.1 ‰.

Isotopic compositions of oxygen and hydrogen were measured on water samples, collected in 12 to 15 mL tightly closed tubes in the field at spring mouths. The $\delta D$ and $\delta^{18}O$ of the water samples, expressed in per mil relative to Vienna Standard Mean Ocean Water (V-SMOW), were measured after Cr reduction and $CO_2$–water equilibration, respectively, on a Finnigan™ MAT-253 mass spectrometer (Thermo Electron Corp., Germany) at CRPG (Nancy, France) (Gajurel et al. 2006; Evans et al. 2008). Experimental uncertainty was $\pm 2‰$ for $\delta D$ and $\pm 0.2‰$ for $\delta^{18}O$.

In addition, isotopic composition of dissolved and gaseous helium was measured on a few water and gas samples, collected in the field in sealed copper tubes and in pre-evacuated stainless steel tubes fitted with valves, respectively. Samples were analyzed at IPGP (Paris, France) using an ultra-high-vacuum purifying glass line linked to an ARESIBO I mass spectrometer (Girault et al. 2014a). Results are given as helium isotope ratio $R/R_A$, defined by the $^4He/^3He$ ratio of the sample divided by the $^4He/^3He$ ratio of the atmosphere $(1.39 \times 10^{-6})$. Experimental uncertainties of $^4He$ content and $R/R_A$ were $\pm 10\%$ and $\pm 0.01$, respectively.

## 8.3. Summary of results at the hydrothermal sites in the Nepal Himalaya

In this section, we present the main results of crustal fluid occurrences at the known hydrothermal systems in the Nepal Himalaya (Figure 8.1). For each Nepal region, we list the sites (with their associated number #) and describe briefly the contexts before giving the most important results at each site. Our review includes data obtained over the last 45 years, from 1975 to 2021.

We focus particularly here only on the hydrothermal sites that have been studied by the authors for the last 23 years, from 1999 to 2021. Some sites are still in natural conditions while others are heavily modified and constructed. Some springs are not thermal springs, but were selected based on their particular interest and location. First, we describe sites in the vicinity of the MCT zone in the different Nepal regions. Then, we describe sites in the vicinity of the MFT zone. Photographs of selected hydrothermal sites are displayed in Figures 8.2 and 8.3. For each site, its location is shown on the map (Figure 8.1) and the mean characteristics of thermal spring water(s) and gaseous emission zone(s) are summarized in the supplementary material (HydroNepal.pdf available at www.iste.co.uk/cattin/himalaya3.zip).

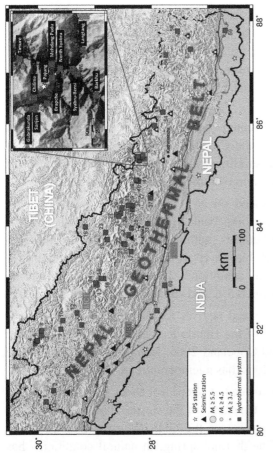

**Figure 8.1.** Map of the Nepal geothermal belt. The currently identified geothermal sites are shown as red squares; their main manifestation is the occurrence of thermal spring(s) and gaseous $CO_2$ and radon emissions (see supplementary material available at www.iste.co.uk/cattin/himalaya3.zip). The main Himalayan thrusts are shown: Main Central Thrust (MCT), Main Boundary Thrust (MBT) and Main Frontal Thrust (MFT). GPS stations (DMG/NSC (Nepal) – UNAVCO (USA)) and seismic stations (DMG/NSC (Nepal) – DASE (France)) are components of international and national networks. Earthquakes from the 1994–2014 catalogue (Adhikari 2021) are also shown with circle size scaled with local magnitude. The inset shows in greater detail the geothermal sites in the Trisuli valley, Central Nepal. Currently known geothermal systems are located in the MCT zone in Far-Western Nepal (1–8), Mid-Western Nepal (9–16), Western Nepal (17–37), Central Nepal (38–55), Eastern Nepal (56–58) and in the MFT zone (59–61). For a color version of this figure, see www.iste.co.uk/cattin/himalaya1.zip

### 8.3.1. *Overview of hydrothermal sites in Far-Western Nepal*

To date, only few hydrothermal sites have been studied in Far-Western Nepal in the Mahakali and Chamliya regions (#1–2) in July 2017 and in the Humla Karnali region (#3–8) in November 2021. However, several other sites certainly exist in this almost unexplored region in terms of hydrothermal systems.

– The Sribagar site (#1) is located at river level about 3 km north of Darchula, in the westernmost part of Nepal along the India–Nepal border, on the eastern bank of the Mahakali river. The site is within rocks of the Greater Himalayan Sequence (GHS) (Robinson et al. 2006). The thermal spring has a maximum temperature of $64.03\pm0.01°C$, the pervasive smell of $H_2S$ is strong and bubbles are present in water.

– The Tapoban site (#2) is located on an alluvial terrace about 25 km northeast of Gokuleshwor, along the Chamliya river, within intercalation of dolomite and phyllite of the Lesser Himalayan Sequence (LHS) (Robinson et al. 2006). The thermal spring has a maximum temperature of $43.02\pm0.01°C$.

– The Kermi site (#3) is located on an alluvial terrace about 14.5 km northwest of Simikot, on the northern bank of the Humla Karnali river, at the confluence of two of its tributaries, within anatectic gneiss of the GHS (Murphy and Copeland 2005; Robinson et al. 2006). Three thermal springs were found almost all in natural conditions. The warmest thermal spring (KER2) has a temperature of $76.9\pm0.5°C$ and shows intense bubbling.

– The Chahara site (#4) is located at river level about 10 km northwest of Simikot, on the western bank of the Humla Karnali river, within anatectic gneiss of the GHS (Murphy and Copeland 2005; Robinson et al. 2006). The thermal spring, in natural conditions, has a temperature of $46.9\pm0.5°C$.

– The Ghadapori site (#5) is located at river level about 9 km northwest of Simikot and 1.4 km southeast of Chahara, on the western bank of the Humla Karnali river, within anatectic gneiss of the GHS (Murphy and Copeland 2005; Robinson et al. 2006). The thermal spring, in natural conditions, has a temperature of $65.8\pm0.2°C$ and a flowrate of $0.27\pm0.03$ L.s$^{-1}$.

– The Kholsi site (#6) is located at river level about 7.5 km northwest of Simikot and 1.5 km southeast of Ghadapori, on the western bank of the Humla Karnali river, within anatectic gneiss of the GHS (Murphy and Copeland 2005; Robinson et al. 2006). The thermal spring, in natural conditions, has a temperature of $67.7\pm0.2°C$.

– The Kharpunath site (#7) is located at river level about 5 km southeast of Simikot, at the confluence of the Chuwa and Humla Karnali rivers, on the northern bank, within gneiss of the GHS (Murphy and Copeland 2005; Robinson et al. 2006). Numerous thermal springs are present near the river or below the river level. The three main thermal springs are arranged with small cemented basins by the local people for bathing. The warmest spring (KHA2) has a temperature of $63.72\pm0.02°C$. A strong pervasive smell of H2S was detected above each cemented basin. Gaseous $CO_2$ emission is compatible with background continental values (Girault 2011).

– The Dojam site (#8) is located at a hillslope about 13.5 km east of Simikot and 3 km southeast of Dojam, on the northern bank of the Lurupya river, a tributary of the Chuwa river, within two-mica gneiss of the GHS (Murphy and Copeland 2005; Robinson et al. 2006). The spring, with red deposits and in natural conditions, has a temperature of $15.17\pm0.01°C$. Gaseous $CO_2$ emission is compatible with background values.

### 8.3.2. *Overview of hydrothermal sites in Mid-Western Nepal*

Several hydrothermal systems were studied in remote Mid-Western Nepal in three main areas: the Rara lake region in Humla in November 2018 (#9–10), the Jumla region in November 2015 (#11–12) and the Lower Dolpo region in November 2008, 2009 and 2010 (#13–16). However, the region is vast, still largely unexplored, and numerous unidentified thermal springs and gaseous emission zones may be present.

– The Soru site (#9) is located at river level about 18 km west of Gamgadhi, on the northern bank of the Mugu Karnali river, within calcareous rocks of the LHS (Iaccarino et al. 2017). The main thermal spring (SORU1), with a temperature of $46.59\pm0.01°C$ and a flowrate of $0.248\pm0.004$ $L.s^{-1}$, has been arranged by local people with a small cemented basin. Observed gaseous $CO_2$ emission is compatible with background values.

– The Nigale site (#10) is located at river level about 2 km north of Gamgadhi, on the northern bank of the Mugu Karnali river, within Kuncha schist and quartzite of the LHS (Iaccarino et al. 2017). A basin was built with dry rocks. The main hot spring (NIG1) has a temperature of $31.3\pm0.1°C$ and a flowrate of $2.0\pm0.5$ $L.s^{-1}$. Observed gaseous $CO_2$ emission remains small.

– The Jumla-Tatopani site (#11) is located on an alluvial terrace at Tatopani, about 12 km southwest of Jumla, on the northern bank of the Tila river, within gneiss of the GHS (Frassi 2015). The main thermal spring (TJ3A), stored in tanks, channeled in pipes flowing into several basins, is used by local people and tourists and has a temperature of $41.9\pm0.1°C$ and a flowrate of $0.79\pm0.01$ $L.s^{-1}$.

– The Lum site (#12) is located on an altitude barren hill about 12 km northeast of Jumla, on the eastern bank of a tributary of the Tila river, within carbonates rocks of the LHS (Frassi 2015). Two springs with different characteristics are channeled and fill a dry rock basin used by the local people for bathing. Large travertine deposits are present. The coolest spring (TJ4A) has a temperature of $15.6\pm0.1°C$; the warmest spring (TJ4B) has a temperature of $24.0\pm0.1°C$. Moderate gaseous $CO_2$ emission is observed.

– The Sulighad site (#13) (Girault et al. 2018b) is located in a mixed forest at river level about 4 km southeast of Juphal, on the western bank of the Rupa Ghad river, a tributary of the Thuli Bheri river, within black mica-schist, schist and marble of the LHS (Carosi et al. 2002; Girault et al. 2012). The hot spring (DLP1), slightly arranged by the local people for bathing, has a temperature of $41.7\pm0.1°C$ and a flowrate of $0.4\pm0.1$ $L.s^{-1}$. Moderate gaseous $CO_2$ emission is observed.

– The DLP2 site (#14) (Girault et al. 2018b) is located in a rockslide at river level about 7 km east of Juphal, on the western bank of the Suli Ghad River, within mica-schist, quartzite and amphibolite of the LHS (Carosi et al. 2002; Girault et al. 2012). The hot spring (DLP2) has a temperature above 25°C and shows bubbling, but $CO_2$ emission remains small.

– The Chhedhul site (#15) (Girault et al. 2018b) is located on a vast alluvial terrace at river level about 15 km east of Dunai, on the southern bank of the Barbung river, near the South Tibetan Detachment System (STDS), within migmatite and gneiss of the GHS and metamorphic carbonates of the Tethyan Sedimentary Sequence (TSS) (Carosi et al. 2002; Girault et al. 2012). The spring (DLP4) has a temperature of $15.4\pm0.1°C$. Surface $CO_2$ fluxes around the springs remain at background levels.

– The Tarakot site (#16) (Girault et al. 2018b) is located in a mixed forest at river level about 13 km southeast of Dunai, on the western bank of the Ghatta river, within black mica-schist, schist and marble of the LHS (Carosi et al. 2002; Girault et al. 2012). The hot spring (DLP3), arranged with a cemented basin by the local people for bathing, has a temperature of $45\pm2°C$ and a flowrate of $0.40\pm0.06$ $L.s^{-1}$. Moderate gaseous $CO_2$ emission is observed.

**Figure 8.2.** *Pictures of selected hydrothermal sites in Far-Western, Mid-Western and Western Nepal. Numbers refer to sites whose location is plotted in Figure 8.1, and characteristics are summarzsed in the supplementary material available at www. iste.co.uk/cattin/himalaya3.zip. For a color version of this figure, see www.iste.co.uk/ cattin/himalaya1.zip*

### 8.3.3. *Overview of hydrothermal sites in Western Nepal*

In the well-known Western Nepal, several hydrothermal systems were studied, including the famous sites located on trekking roads, in eight main areas: the Myagdi region (#17) in November 2010, the Kali Gandaki region (#19–23) in December 2009, the Modi region (#24) in July 2021, the Seti region (#25) in November 2010, the Pokhara region (#31) in May and November 2016 and May and July 2017, the Khudi region (#32) in January 2017 and 2018, the Marsyandi region (#26–30 and #33–36) in December 2010, May and November 2016, January 2017, January and August 2018, April and August 2019 and the Gorkha region (#37) in May 2016.

– The Beni site (#17) (Grabczak and Kotarba 1985; Sharma 1995; Evans et al. 2004, 2008; Girault et al. 2014b) is located on an alluvial terrace about 7 km northwest of Beni Bazar, on the northern bank of the Myagdi river, within alternation of Kuncha phyllite and quartzite of the LHS (Upreti and Yoshida 2005; Girault et al. 2012). The hot spring (BEN), located in a closed compound, stored in tanks and channelized to fill a large cemented swimming pool for bathing, has a temperature of $49.78\pm0.04°C$ and a flowrate of $2.0\pm0.2$ $L.s^{-1}$. Observed gaseous $CO_2$ emission is compatible with background values.

– The Ratopani site (#20) (Evans et al. 2004, 2008) is located a few meters above river level about 17 km northeast of Beni Bazar, on the eastern bank of the Kali Gandaki river, within Kuncha phyllite and quartzite of the LHS (Upreti and Yoshida 2005; Girault et al. 2012). The hot spring, stored in tanks and filling cemented basins used by local people for bathing, has a temperature of $51.38\pm0.04°C$ and a flowrate of $0.0020\pm0.0001$ $L.s^{-1}$.

– The Tatopani site (#21) (Grabczak and Kotarba 1985; Bhattarai 1986; Evans et al. 2004, 2008; Girault et al. 2014b) is located at river level on an alluvial terrace about 19 km northeast of Beni Bazar, on the western bank of the Kali Gandaki river, within garnet-schist and quartzite of the LHS (Upreti and Yoshida 2005; Girault et al. 2012). The well-known hot spring (KAL1), pumped near surface and filling several cemented basins used by local people and visitors for bathing, has a temperature of $60.07\pm0.01°C$ and a flowrate of $1.8\pm0.1$ $L.s^{-1}$. Moderate gaseous $CO_2$ emission is observed.

– The Narchyeng site (#22) (Grabczak and Kotarba 1985; Girault et al. 2014b) is located at river level and on an alluvial terrace about 20 km northeast of Beni Bazar and 500 m north of Tatopani site, on the eastern bank of the Kali Gandaki river, within garnet-schist and quartzite of the LHS (Upreti and Yoshida 2005; Girault et al. 2012). The hot spring has a temperature of

$59.86\pm0.02°C$ and a flowrate of $0.80\pm0.03$ $L.s^{-1}$. Moderate gaseous $CO_2$ emission is observed.

– The Jhinudanda site (#24) (Grabczak and Kotarba 1985; Bhattarai 1986; Evans et al. 2004, 2008) is located at river level about 13 km northeast of Birethanti, on the western bank of the Modi river, within mica-schist of the LHS (Pearson and DeCelles 2005). The main hot spring (JND2), slightly arranged by the local people with cemented basins (regularly destroyed by floods), has a temperature of $39.02\pm0.01°C$.

– The Seti site (#25) (Grabczak and Kotarba 1985; Evans et al. 2004, 2008; Girault et al. 2014b) is located at river level about 16 km northeast of Pokhara, on the eastern and western banks of the Seti river, within quartzite of the LHS. The hot spring, arranged in dry stone basins and used by many local people, has a temperature of $43.02\pm0.05°C$ and a flowrate of $0.30\pm0.02$ $L.s^{-1}$. Moderate gaseous $CO_2$ emission is observed.

– The Chame site (#28) (Becker 2005; Becker et al. 2008; Ghezzi et al. 2019) is located on an alluvial terrace about 200 m north of Chame, on the northern bank of the Marsyandi river, within gneiss of the GHS (Colchen et al. 1986). The warmest hot spring (TCHM1A), arranged with a large cemented basin and pipes, has a temperature of $51.68\pm0.06°C$. The coldest hot spring (TCHM1B) has a temperature of $22.40\pm0.07°C$ and a flowrate of $8.5\pm0.1$ $L.s^{-1}$.

– The Dhor Barahi site (#31) (Perrier et al. 1999, 2002b) is located on a forested hillslope about 16 km southeast of Pokhara, on the western bank of the Seti river, within dolomite of the LHS. Located in the compound of a Hindu temple, the spring is periodic, with repetition rate of about 30 min, has a temperature of $22.0\pm0.1°C$ and an approximate flowrate of $17\pm5$ $L.s^{-1}$.

– The Probi site (#32) (Girault et al. 2018c) is located at river level at the base of a large outcrop about 7.5 km north of Khudi Bazar, on the western bank of the Khudi river, within alternation of schist and quartzite of the LHS (Pearson and DeCelles 2005; Thapa et al. 2018, 2019; Tamang et al. 2019). The site has remained almost untouched. The hot spring with bubbling (PRO2) has a temperature of $30.1\pm0.1°C$. The hot spring with the largest flowrate (PRO4) has a temperature of $24.1\pm0.1°C$ and a flowrate of $1.0\pm0.3$ $L.s^{-1}$. Strong gaseous $CO_2$ emission is observed.

– The Jagat site (#33) (Evans et al. 2004, 2008; Becker 2005; Becker et al. 2008; Girault et al. 2018c; Ghezzi et al. 2019) is located at river level between 1.5 and 2.5 km north of Shirchaur, on the western bank of the Marsyandi river, within gneiss of the GHS (Colchen et al. 1986; Girault et al. 2012). To the south, the hot spring (JAG1), arranged into cemented basins by the local people for bathing, has a temperature of 54.68±0.04°C. Observed gaseous $CO_2$ emission is compatible with background values. To the north, the warmest hot spring (JAG2A), almost untouched, has a temperature of 52.20±0.06°C.

– The Shirchaur site (#34) (Becker 2005; Becker et al. 2008; Girault et al. 2018c) is located at the base of a mixed forest a few meters above river level about 6.5 km north of the main Bahundanda hot springs, on the eastern bank of the Marsyandi valley, within gneiss and mica-schist of the GHS (Girault et al. 2012; Thapa et al. 2018, 2019; Tamang et al. 2019). The hot spring, channelized into a pipe, has a temperature of 34.00±0.07°C and a flowrate of 0.115±0.002 L.s$^{-1}$.

– The Bahundanda site (#35–36) (Evans et al. 2004, 2008; Becker 2005; Becker et al. 2008; Girault et al. 2014b, 2018c) is located on several alluvial and debris fall deposit terraces about 12 km north of Besisahar on the eastern bank of the Marsyandi river within alternation of schist and quartzite of the LHS (Pearson and DeCelles 2005; Girault et al. 2012; Thapa et al. 2018, 2019; Tamang et al. 2019). The site comprises several bubbling hot springs and gaseous emission zones. In the north (#35), the main hot spring (MARS2), slightly arranged with cemented basins, has a temperature of 48.05±0.01°C and a flowrate of 0.325±0.006 L.s$^{-1}$. On the alluvial terrace, about 330 m to the south, the main hot spring (MARS4B) has a temperature of 32.72±0.01°C. About 300 m to the south again (#36), another spring (MARS8B) has a temperature of 23.10±0.07°C. Strong gaseous $CO_2$ emission is observed.

– The Bhulbule site (#37) (Evans et al. 2004, 2008; Girault et al. 2018c) is located on a cultivated alluvial plain about 8 km east of Gorkha, near the Daraundi river, within rocks of the LHS (Dhital 2015). The hot spring, arranged in a cemented basin by local people, has a temperature of 34.76±0.09°C.

### 8.3.4. *Overview of hydrothermal sites in Central Nepal*

In Central Nepal, numerous hydrothermal systems were studied in four main areas: the Budhi Gandaki region (#38–41) in January 2017 and 2018, the Upper Trisuli region encompassing the Bhote Kosi (#46, 48–51), Chilime (#42–45, 47) and Langtang valleys (#52) with a large number of fieldworks from 2004 to 2021, the Bhote Koshi region (#53) in August 2010, January 2011, May 2016, December 2018 and January 2021 and the Tama Koshi region

(#54–55) in May 2019. The Upper Trisuli valley remains the most studied region in Nepal so far in terms of hydrothermal systems.

– The Tatopani site (#38) (Evans et al. 2004, 2008; Girault et al. 2018c) is located at the base of a large outcrop about 21 km north of Arkhet Bazar and 8.5 km south of Jagat, on the western bank of the Budhi Gandaki river, within quartzite and augen gneiss of the LHS (Khanal and Robinson 2013). The main hot spring (BUD4C), channelized in a pipe discharging in a cemented basin used by the local people for washing, has a temperature of 49.94±0.07°C and a flowrate of 0.061±0.001 L.s$^{-1}$. Another spring (BUD4B), discharging at the same place, has a temperature of 24.01±0.09°C and a similar flowrate of 0.060±0.001 L.s$^{-1}$.

– The Lysiapu site (#39) (Evans et al. 2004, 2008; Girault et al. 2018c) is located on an alluvial terrace about 20 km north of Arkhet Bazar and 700 m southwest of Tatopani, on the western bank of the Budhi Gandaki river, within garnet-rich mica-schist and quartzite of the LHS (Khanal and Robinson 2013). The northern hot spring (BUDX1), located on the trekking path, has a temperature of 25.1±0.1°C and a flowrate of 1.0±0.2 L.s$^{-1}$. The southern hot spring (BUD2), located near the river, has a temperature of 52.32±0.09°C.

– The Khorlabesi site (#40) (Girault et al. 2018c) is located a few meters above river level about 17.5 km north of Arkhet Bazar and 2.5 m north of Machhakhola, on the western bank of the Budhi Gandaki river, within garnet-rich mica-schist of the LHS (Khanal and Robinson 2013). The hot spring shows a large travertine dome, has a temperature of 31.6±0.1°C and a flowrate of 0.6±0.2 L.s$^{-1}$. Along the trekking path south of the spring, where pervasive smell of $H_2S$ is detected, significant gaseous $CO_2$ emission is observed.

– The Machhakhola site (#41) (Girault et al. 2018c) is located at river level about 20 km north of Arkhet Bazar and 700 m southwest of Tatopani, on both banks of the Budhi Gandaki river, within mica-schist of the LHS (Khanal and Robinson 2013). The main hot spring (BUD0), on the eastern bank with large bubbling, has a temperature of 59.75±0.07°C. On the western bank, another hot spring (BUD6) has a temperature of 57.1±0.1°C. Around BUD0 hot spring, strong gaseous $CO_2$ emission is observed.

– The Gubrakunda site (#42) (Girault et al. 2018c) is located in an altitude mixed forest about 4.5 km northwest of Chilime, on the western bank of a tributary of the Chilime river, within alternation of dolomite and graphitic mica-schist of the LHS (Girault et al. 2012; Dhital 2015). This site is characterized by two springs, considered as sacred by the local people. One hot spring (GUB1), filling a large bubbling pond, has a temperature of 15.85±0.01°C. The other hot spring (GUB2), forming a large prominent travertine, has a temperature of 13.31±0.01°C and a flowrate of 2.06±0.06 L.s$^{-1}$. Around GUB2 spring, significant gaseous $CO_2$ emission is observed.

– The Sanjen site (#43) (Girault et al. 2018c) is located on an alluvial terrace about 4 km northwest of Chilime, on the western bank of the Sanjen river, a tributary of the Chilime river, within alternation of dolomite and graphitic mica-schist of the LHS (Girault et al. 2012; Dhital 2015). This site is a hydroelectric project under construction with boreholes and tunnels showing bubbling and $CO_2$ release. The hydrothermal aquifer (DH1 and DH2), sampled from the two boreholes from where a strong pervasive smell of $H_2S$ is detected, has a mean temperature of 19.9±0.2°C. The discharged water in the lower (TSJ3) and upper (TSJ5B) tunnels has a temperature of 20.3±0.1°C and 36.90±0.07°C, respectively. Strong gaseous $CO_2$ emission is observed inside the lower tunnel.

– The Chilime site (#44) (Le Fort 1975; Grabczak and Kotarba 1985; Bhattarai 1986; Perrier et al. 2002a, 2009; Evans et al. 2004, 2008; Becker 2005; Becker et al. 2008; Girault et al. 2014b, 2018b, c) is located in an altitude mixed forest at the Paragaon village about 4 km north of Chilime, on the eastern bank of the Chilime river, within graphitic mica-schist and dolomite of the LHS (Girault et al. 2012; Dhital 2015). The main hot spring (CHI), a major pilgrimage site arranged with large shallow cemented basins for bathing, has a temperature of 48.10±0.03°C and a flowrate of 4.8±0.1 L.s$^{-1}$. Strong gaseous $CO_2$ emission is observed above the spring. The hot spring and $CO_2$ emissions were dramatically modified by the 2015 Gorkha earthquake (Girault et al. 2018c).

– The Brapche site (#45) (Girault et al. 2018c) is located at river level about 2 km northwest of Chilime, on the western bank of the Chilime river, within quartzite and amphibolite of the LHS (Girault et al. 2012; Dhital 2015). The hot spring, slightly arranged with a dry stone basin by the local people, has a temperature of 44.10±0.01°C and a flowrate of 0.80±0.02 L.s$^{-1}$.

– The Barkhu site (#46) (Girault et al. 2018c) is located at river level about 2.5 km north of Dhunche, on the eastern bank of the Trisuli river, within Kuncha schist of the LHS (Girault et al. 2012; Dhital 2015). The hot spring, slightly arranged by the local people in natural basins for bathing, has a temperature of 49.9±0.1°C. Observed gaseous $CO_2$ emission is compatible with background values.

– The Pajung site (#47) is located on a hillslope about 3 km east of Chilime and 3 km north of Syabru-Bensi, on the northern bank of the Chilime river, within mica-schist and augen gneiss of the LHS (Girault et al. 2012; Dhital 2015). The main hot spring (PAJ2), still in natural conditions, has a temperature of 18.32±0.01°C. Around the springs and the cavities where a strong pervasive smell of $H_2S$ is present, strong gaseous $CO_2$ emission is observed.

– The Syabru-Bensi site (#48) (Le Fort 1975; Grabczak and Kotarba 1985; Perrier et al. 2002a, 2009; Evans et al. 2004, 2008; Becker 2005; Becker et al. 2008; Byrdina et al. 2009; Girault et al. 2009; Girault and Perrier 2014; Girault et al. 2014a, b, 2018b, c), located on alluvial terraces about 7 km northeast of Dhunche, on both banks of the Trisuli river, within mica-schist, quartzite and augen gneiss of the LHS (Girault et al. 2012; Dhital 2015), is one of the most studied sites in Nepal. It exhibits a large number of hot springs; representatively, the two main hot springs (SBP0 and SBB5), located on the western bank and arranged with cemented basins by local people for bathing, have different characteristics. This site was modified by the 2015 Gorkha earthquake (Girault et al. 2018). After 2015, the main hot spring (SBP0) has a temperature of 66.470±0.004°C and a flowrate of 0.0515±0.0003 $L.s^{-1}$, and the second spring (SBB5) has a temperature of 31.680±0.001°C and a flowrate of 0.124±0.001 $L.s^{-1}$. Other seepages are also observed in numerous places around the main springs and a large gaseous $CO_2$ emission is detected up to 500 m from the springs.

– The North Syabru site (#49) (Evans et al. 2004, 2008; Becker 2005; Becker et al. 2008; Byrdina et al. 2009; Girault et al. 2014b, 2018c) is located at river level about 2.3 km north of Syabru-Bensi, on the western bank of the Bhote Kosi river, within augen gneiss of the LHS (Girault et al. 2012; Dhital 2015). The hot spring, remaining in natural conditions and precipitating red deposits, has a temperature of 24.531±0.002°C and shows significant bubbling.

– The Mehdang Pudu site (#50) (Girault et al. 2018c) is located at river level about 4 km northwest of Syabru-Bensi and 6.5 km south of Timure, on the eastern bank of the Bhote Kosi river, within quartzite of the GHS (Girault et al. 2012; Dhital 2015). In the north, the warmest hot spring (MEH1) has a temperature of 62.41±0.01°C. In the south, another hot spring (MEH2) has a temperature of $58.88\pm0.01°C$ and a flowrate of $0.246\pm0.003$ L.s$^{-1}$. MEH1 spring is bubbling and high $CO_2$ concentration (>1.5%) is detected on the ground surface about 20 m above the hot springs.

– The Timure site (#51) (Evans et al. 2004, 2008; Becker 2005; Becker et al. 2008; Girault et al. 2014b, 2018b, c) is located on a large alluvial terrace about 9 km north of Syabru-Bensi and 1.5 km south of Timure, on the eastern bank of the Bhote Kosi river, within quartzite of the GHS (Girault et al. 2012; Dhital 2015). After 2017, the main hot spring (TIMN), with increasing importance in the valley, was arranged with large cemented basins by local people for bathing. TIMN spring has a mean temperature of $62.910\pm0.004°C$. The site exhibits also two significant gaseous $CO_2$ emission zones, one in the vicinity of the main hot spring and the other up to 500 m away from it.

– The Langtang Pahiro site (#52) (Evans et al. 2004, 2008; Girault et al. 2014b, 2018b, c) is located on an alluvial terrace in a dense forest about 3.7 km east of Syabru-Bensi on the northern bank of the Langtang river within gneiss, mica-schist and quartzite of the GHS (Girault et al. 2012; Dhital 2015). After 2015, the main hot spring (LPAH), whose only access is by crossing in the Langtang river (no bridge), is slightly arranged by local people for bathing and has a temperature of $37\pm1°C$. Gaseous $CO_2$ emission is compatible with background values.

– The Kodari site (#53) (Sharma 1995; Perrier et al. 2002a; Evans et al. 2004, 2008; Girault et al. 2014b, 2018c) is located on an alluvial terrace about 18 km north of Bahrabise and 3 km south of Kodari, on the western bank of the Bhote Koshi river, within quartzite and augen gneiss of the LHS (Rai and Upreti 2006 Unpublished; Girault et al. 2012). After 2015, the main hot spring (KOD), a famous pilgrimage site arranged with several basins and related buildings, has a temperature of $50.21\pm0.01°C$ and a flowrate of $3.00\pm0.04$ L.s$^{-1}$. Gaseous $CO_2$ emission is compatible with background values.

– The Tatopani (Tama K.) site (#54–55) is located on a debris fall deposit at river level about 8.5 km northeast of Singati and 5 km south of Gongar, on the western bank of the Tama Koshi river, within carbonates and phyllite of the LHS (Dhital 2015). In the north, the main spring (TAM1), arranged in a small pond by local people, has a temperature of $32.12\pm0.04°C$. In the south, another hot spring (TAM2A), in natural conditions, has a temperature of $41.73\pm0.01°C$. Moderate gaseous $CO_2$ emission is observed.

**Figure 8.3.** *Pictures of selected hydrothermal sites in Central and Eastern Nepal and in the MFT zone. Numbers refer to sites whose location is plotted in Figure 8.1, and characteristics are summarized in supplementary material available at www.iste.co. uk/cattin/himalaya3.zip. For a color version of this figure, see www.iste.co.uk/cattin/ himalaya1.zip*

### 8.3.5. *Overview of hydrothermal sites in Eastern Nepal*

In Eastern Nepal, only few hydrothermal systems were studied in three main areas: the Arun region (#56) in December 2020, the Piluwa region (#57) in October 2021 and the Tamor region (#58) in October 2021. Before 2016, no hot spring was reported from this region. The region is vast and numerous unidentified thermal springs and gaseous emission zones may be present.

– The Hatiya site (#56) is located on an alluvial terrace about 42 km north of Khandbari, on the western bank of the Arun river, within gneiss of the GHS (Lombardo et al. 1993; Dhital 2015). Three hot springs were found at a place arranged with cemented basins and pipes by local people. The warmest hot spring (HTA2) has a temperature of $37.05\pm0.01°C$ and a flowrate of $0.0279\pm0.0002$ $L.s^{-1}$. The hot spring with the highest flowrate (HTA1) has a temperature of $35.68\pm0.01°C$ and a flowrate of $0.13\pm0.01$ $L.s^{-1}$. Gaseous $CO_2$ emission is compatible with background values.

– The Nundhaki site (#57) is located on a vegetated hillslope about 25 km southeast of Khandbari, on the southern bank of the Piluwa river, a tributary of the Arun river, within gneiss of the GHS (Lombardo et al. 1993; Dhital 2015). The spring, channelized between cemented walls, has a temperature of $18.26\pm0.03°C$. Gaseous $CO_2$ emission is compatible with background values.

– The Sekhathum site (#58) is located on a debris fall deposit about 35 km northeast of Taplejung, on the eastern bank of the Tamor river, within gneiss of the GHS (Shrestha et al. 1984; Dhital 2015). The spring was burried by a rockslide and could not be sampled in 2021.

### 8.3.6. *Overview of hydrothermal sites in the MFT zone*

In the MFT zone, only a few hydrothermal systems were studied in three main areas: the Satakshidam valley (#59) in Eastern Nepal in December 2020 and the Surai (#60) and Rapti regions (#61) in Mid-Western Nepal in November 2016. Probably numerous other thermal springs, similar or smaller, are unidentified in the MFT and MBT zones throughout Nepal.

– The Satakshidam site (#59) is located at river level about 55 km east of Dharan and 5 km north of Satakshidam within Middle and Lower Siwaliks (Dhital 2015, and references herein). The spring has a temperature of $21.972\pm0.004°C$. Gaseous $CO_2$ emission is small and not significant.

– The Surainaka site (#60) (Bhattarai 1986; Evans et al. 2004) is located in an incised valley at river level about 62 km west of Butwal and 12 km north of Shivapur, on the banks of the Surai river, within sandstone and siltstone of the Middle Siwaliks (Dhital 2015, and references herein). Four hot springs were found almost untouched. The main hot spring in the south (SUR1) has a temperature of $33.8\pm0.1$°C and a flowrate of $0.250\pm0.003$ L.s$^{-1}$. The main hot spring in the north (SUR3) has a temperature of $27.3\pm0.1$°C and a flowrate of $0.0303\pm0.0001$ L.s$^{-1}$. Gaseous $CO_2$ emission is small and not significant.

– The Rihar site (#61) (Bhattarai 1986) is located on alluvial sediments about 72 km east of Nepalganj and 16 km west of Lamahi, along the east–west road, near the Rapti river, within Upper Siwaliks (Dhital 2015, and references herein). The warmest spring (RIH1), pumped by local people, has a temperature of $36.5\pm0.1$°C and a flowrate of $1.50\pm0.09$ L.s$^{-1}$. The main spring in the village (RIH3), collected in a tank, has a temperature of $32.8\pm0.1$°C. Gaseous $CO_2$ emission is small and not significant.

## 8.4. Conclusion

In this chapter, we have presented an overview of the currently available information and data on hydrothermal systems in Nepal. While more than 60 sites are now studied, numerous sites remain to be discovered in vast unexplored regions. Although a larger number of sites (n = 18) have been studied in detail in Central Nepal, with particular emphasis in the Trisuli valley, most of the well-investigated sites are located in the vicinity of the MCT shear zone, defining the "Nepal geothermal belt". Thermal springs are the most frequently identified surface manifestations, and some of them, with sufficient flowrate to sustain economic value, are pilgrimage sites for the local population, and even sometimes for the inhabitants of bigger cities of Kathmandu and Pokhara. Numerous thermal springs are located in remote areas in the MCT zone, as well as in the MBT and MFT zones, which are almost unexplored in terms of hydrothermal systems.

In Volume 3 – Chapter 9, we rely on this unprecedented dataset to present the current knowledge on hydrothermal systems in the Nepal Himalaya and their sensitivity to the Himalayan earthquake cycle.

## 8.5. References

Ader, T., Avouac, J.-P., Liu-Zeng, J., Lyon-Caen, H., Bollinger, L., Galetzka, J., Genrich, J., Thomas, M., Chanard, K., Sapkota, S.N. et al. (2012). Convergence rate across the Nepal Himalaya and interseismic coupling on the Main Himalayan Thrust: Implications for seismic hazard. *J. Geophys. Res.*, 117, B04403.

Adhikari, L.B. (2021). Seismicity associated with the April 25, 2015 Gorkha earthquake in Nepal: Probing the Himalayan Seismic Cycle. Université de Paris, Institut de Physique du Globe de Paris, Paris.

Adhikari, L.B., Gautam, U.P., Koirala, B.P., Bhattarai, M., Kandel, T., Gupta, R.M., Timsina, C., Maharjan, N., Maharjan, K., Dahal, T. et al. (2015). The aftershock sequence of the April 25th 2015 Gorkha-Nepal earthquake. *Geophys. J. Int.*, 203, 2119–2124.

Adhikari, L.B., Bollinger, L., Vergne, J., Lambotte, S., Chanard, K., Laporte, M., Li, L., Koirala, B.P., Bhattarai, M., Timsina, C. et al. (2021). Orogenic collapse and stress adjustments revealed by an intense seismic swarm following the 2015 Gorkha earthquake in Nepal. *Front. Earth Sci.*, 9, 659937.

Ague, J.J. (2000). Release of $CO_2$ from carbonate rocks during regional metamorphism of lithologically heterogeneous crust. *Geology*, 28, 1123–1126.

Assayag, N., Rivé, K., Ader, M., Jézéquel, D., Agrinier, P. (2006). Improved method for isotopic and quantitative analysis of dissolved inorganic carbon in natural water samples. *Rapid Commun. Mass Spectrom.*, 20, 2243–2251.

Avouac, J.-P. (2003). Mountain building, erosion, and the seismic cycle in the Nepal Himalaya. *Adv. Geophys.*, 46, 1–80.

Avouac, J.-P., Bollinger, L., Lavé, J., Cattin, R., Flouzat, M. (2001). Le cycle sismique en Himalaya. *C. R. Acad. Sci.*, 333, 513–529.

Becker, J.A. (2005). Quantification of Himalayan metamorphic $CO_2$ fluxes: Impact on global carbon budgets. Corpus Christi College. University of Cambridge, Cambridge.

Becker, J.A., Bickle, M.J., Galy, A., Holland, T.J.B. (2008). Himalayan metamorphic $CO_2$ fluxes: Quantitative constraints from hydrothermal springs. *Earth Planet. Sci. Lett.*, 265, 616–629.

Bettinelli, P., Avouac, J.-P., Flouzat, M., Bollinger, L., Ramillien, G., Rajaure, S., Sapkota, S. (2008). Seasonal variations of seismicity and geodetic strain in the Himalaya induced by surface hydrology. *Earth Planet. Sci. Lett.*, 266, 332–344.

Bhattarai, D.R. (1980). Some geothermal springs of Nepal. *Tectonophysics*, 62, 7–11.

Bhattarai, D.R. (1986). Geothermal manifestations in Nepal. *Geothermics*, 15, 715–717.

Bilham, R. (2019). Himalayan earthquakes: A review of historical seismicity and early 21st century slip potential. In *Himalayan Tectonics: A Modern Synthesis*, Treloar, P.J. and Searle, M.P. (eds). Geological Society Special Publications, London.

Bollinger, L., Avouac, J.-P., Cattin, R., Pandey, M.R. (2004). Stress buildup in the Himalaya. *J. Geophys. Res.*, 109, B11405.

Burton, M.R., Sawyer, G.M., Granieri, D. (2013). Deep carbon emissions from volcanoes. *Rev. Mineral. Geochem.*, 75, 323–354.

Byrdina, S., Revil, A., Pant, S.R., Koirala, B.P., Shrestha, P.L., Tiwari, D.R., Gautam, U.P., Shrestha, K., Sapkota, S.N., Contraires, S. et al. (2009). Dipolar self-potential anomaly associated with carbon dioxide and radon flux at Syabru-Bensi hot springs in central Nepal. *J. Geophys. Res.*, 114, B10101.

Carosi, R., Montomoli, C., Visonà, D. (2002). Is there any detachment in the Lower Dolpo (western Nepal). *C. R. Geoscience*, 334, 933–940.

Chanard, K., Avouac, J.-P., Ramilien, G., Genrich, J. (2014). Modeling deformation induced by seasonal variations of continental water in the Himalaya region: Sensitivity to Earth elastic structure. *J. Geophys. Res. Solid Earth*, 119, 5097–5113.

Chiodini, G., Cioni, R., Guidi, M., Raco, B., Marini, L. (1998). Soil $CO_2$ flux measurements in volcanic and geothermal areas. *Appl. Geochem.*, 13, 543–552.

Colchen, M., Le Fort, P., Pêcher, A. (1986). *Carte géologique Annapurna-Manaslu-Ganesh, Himalaya du Népal*. CNRS, Paris.

Derry, L.A., Evans, M.J., Darling, R., France-Lanord, C. (2009). Hydrothermal heat flow near the Main Central Thrust, central Nepal Himalaya. *Earth Planet. Sci. Lett.*, 286, 101–109.

Dhital, M.R. (2015). *Geology of the Nepal Himalaya. Regional Perspective of the Classic Collided Orogen*. Springer International Publishing, Cham.

Evans, M.J., Derry, L.A., France-Lanord, C. (2004). Geothermal fluxes of alkalinity in the Narayani river system of central Nepal. *Geochem. Geophys. Geosyst.*, 5, Q08011.

Evans, M.J., Derry, L.A., France-Lanord, C. (2008). Degassing of metamorphic carbon dioxide from the Nepal Himalaya. *Geochem. Geophys. Geosyst.*, 9, Q04021.

Frassi, C. (2015). Dominant simple-shear deformation during peak metamorphism for the lower portion of the Greater Himalayan Sequence in West Nepal: New implications for hybrid channel flow-type mechanisms in the Dolpo region. *J. Struct. Geol.*, 81, 28–44.

Gaillardet, J. and Galy, A. (2008). Himalaya–Carbon sink or source? *Science*, 320, 1727–1728.

Gaillardet, J., Dupré, B., Louvat, P., Allègre, C.J. (1999). Global silicate weathering and $CO_2$ consumption rates deduced from the chemistry of large rivers. *Chem. Geol.*, 159, 3–30.

Gajurel, A.P., France-Lanord, C., Huyghe, P., Guilmette, C., Gurung, D. (2006). C and O isotope compositions of modern fresh-water mollusc shells and river waters from Himalaya and Ganga plain. *Chem. Geol.*, 233, 156–183.

Galetzka, J., Melgar, D., Genrich, J.F., Geng, J., Owen, S., Lindsey, E.O., Xu, X., Bock, Y., Avouac, J.-P., Adhikari, L.B. et al. (2015). Slip pulse and resonance of Kathmandu basin during the 2015 Mw 7.8 Gorkha earthquake, Nepal imaged with geodesy. *Science*, 349, 1091–1095.

Ghezzi, L., Petrini, R., Montomoli, C., Carosi, R., Paudyal, K., Cidu, R. (2017). Findings on water quality in Upper Mustang (Nepal) from a preliminary geochemical and geological survey. *Environ. Earth Sci.*, 76, 651.

Ghezzi, L., Iaccarino, S., Carosi, R., Montomoli, C., Simonetti, M., Paudyal, K.R., Cidu, R., Petrini, R. (2019). Water quality and solute sources in the Marsyangdi River system of Higher Himalayan range (West-Central Nepal). *Sci. Total Environ.*, 677, 580–589.

Girault, F. (2011). Étude des flux de dioxyde de carbone et de radon dans l'Himalaya du Népal (in French). University Paris Diderot (Paris VII), Paris.

Girault, F. and Perrier, F. (2014). The Syabru-Bensi hydrothermal system in central Nepal 2: Modeling and significance of the radon signature. *J. Geophys. Res. Solid Earth*, 119, 4056–4089.

Girault, F., Koirala, B.P., Perrier, F., Richon, P., Rajaure, S. (2009). Persistence of radon-222 flux during monsoon at a geothermal zone in Nepal. *J. Environ. Radioact.*, 100, 955–964.

Girault, F., Perrier, F., Gajurel, A.P., Bhattarai, M., Koirala, B.P., Bollinger, L., Fort, M., France-Lanord, C. (2012). Effective radium concentration across the Main Central Thrust in the Nepal Himalayas. *Geochim. Cosmochim. Acta*, 98, 203–227.

Girault, F., Perrier, F., Crockett, R., Bhattarai, M., Koirala, B.P., France-Lanord, C., Agrinier, P., Ader, M., Fluteau, F., Gréau, C., Moreira, M. (2014a). The Syabru-Bensi hydrothermal system in central Nepal 1: Characterization of carbon dioxide and radon fluxes. *J. Geophys. Res. Solid Earth*, 119, 4017–4055.

Girault, F., Bollinger, L., Bhattarai, M., Koirala, B.P., France-Lanord, C., Rajaure, S., Gaillardet, J., Fort, M., Sapkota, S.N., Perrier, F. (2014b). Large-scale organization of carbon dioxide discharge in the Nepal Himalayas. *Geophys. Res. Lett.*, 41, 6358–6366.

Girault, F., Perrier, F., Przylibski, T.A. (2018a). Radon-222 and radium-226 occurrence in water: A review. *Geol. Soc. Lond. S. P.*, 451, 131–154.

Girault, F., Koirala, B.P., Bhattarai, M., Perrier, F. (2018b). Radon and carbon dioxide around remote Himalayan thermal springs. *Geol. Soc. Lond. S. P.*, 451, 155–181.

Girault, F., Adhikari, L.B., France-Lanord, C., Agrinier, P., Koirala, B.P., Bhattarai, M., Mahat, S.S., Groppo, C., Rolfo, F., Bollinger, L. et al. (2018c). Persistent $CO_2$ emissions and hydrothermal unrest following the 2015 earthquake in Nepal. *Nat. Comm.*, 9, 2956.

Grabczak, J. and Kotarba, M. (1985). Isotopic composition of the thermal waters in the central part of the Nepal Himalayas. *Geothermics*, 14, 567–575.

Iaccarino, S., Montomoli, C., Carosi, R., Massonne, H.-J., Vison, D. (2017). Geology and tectono-metamorphic evolution of the Himalayan metamorphic core: Insights from the Mugu Karnali transect, Western Nepal (Central Himalaya). *J. Metamorph. Geol.*, 35, 301–325.

Ingebritsen, S.E. and Manning, C.E. (2010). Permeability of the continental crust: Dynamic variations inferred from seismicity and metamorphism. *Geofluids*, 10, 193–205.

Irwin, W.P. and Barnes, I. (1980). Tectonic relations of carbon dioxide discharges and earthquakes. *J. Geophys. Res.*, 85, 3115–3121.

Kemeny, P.C., Lopez, G.I., Dalleska, N.F., Torres, M., Burke, A., Bhatt, M.P., West, A.J., Hartmann, J., Adkins, J.F. (2021). Sulfate sulfur isotopes and major ion chemistry reveal that pyrite oxidation counteracts $CO_2$ drawdown from silicate weathering in the Langtang-Trisuli-Narayani River system, Nepal Himalaya. *Geochim. Cosmochim. Acta*, 294, 43–69.

Kerrick, D.M. and Caldeira, K. (1998). Metamorphic $CO_2$ degassing from orogenic belts. *Chem. Geol.*, 145, 213–232.

Khanal, S. and Robinson, D.M. (2013). Upper crustal shortening and forward modeling of the Himalayan thrust belt along the Budhi-Gandaki River, central Nepal. *Int. J. Earth. Sci.*, 102, 1871–1891.

Kotarba, M. (1986). Hydrogeological investigations in Seti Khola and Trisuli thermal springs areas (Nepal Himalayas). *Geologia*, 12, 37–51.

Le Fort, P. (1975). Himalayas: Collided range, present knowledge of continental arc. *Am. J. Sci.*, A275, 1–44.

Lemonnier, C., Marquis, G., Perrier, F., Avouac, J.-P., Chitrakar, G., Kafle, B., Sapkota, S., Gautam, U., Tiwari, D., Bano, M. (1999). Electrical structure of the Himalaya of Central Nepal: High conductivity around the mid-crustal ramp along the MHT. *Geophys. Res. Lett.*, 26, 3261–3264.

Lombardo, B., Pertusati, P., Borghi, S. (1993). Geology and tectonomagmatic evolution of the eastern Himalaya along the Chomolungma-Makalu transect. In *Himalayan Tectonics*, Treloar, P.J. and Searle, M.P. (eds). Geological Society Special Publication, London.

Manga, M., Beresnev, I., Brodsky, E.E., Elkhoury, J.E., Elsworth, D., Ingebritsen, S.E., Mays, D.C., Wang, C.-Y. (2012). Changes in permeability caused by transient stresses: Field observations, experiments, and mechanisms. *Rev. Geophys.*, 50, RG2004.

Manning, C.E. and Ingebritsen, S.E. (1999). Permeability of the continental crust: Implications of geothermal data and metamorphic systems. *Rev. Geophys.*, 37, 127–150.

Märki, L., Lupker, M., France-Lanord, C., Lavé, J., Gallen, S., Gajurel, A.P., Haghipour, N., Leuenberger-West, F., Eglinton, T. (2021). An unshakable carbon budget for the Himalaya. *Nat. Geosci.*, 14, 745–750.

Miller, S.M. (2020). Aftershocks are fluid-driven and decay rates controlled by permeability dynamics. *Nat. Comm.*, 11, 5787.

Mörner, N.-A. and Etiope, G. (2002). Carbon degassing from the lithosphere. *Global Planet. Change*, 33, 185–203.

Murphy, M.A. and Copeland, P. (2005). Transtensional deformation in the central Himalaya and its role in accommodating growth of the Himalayan orogen. *Tectonics*, 24, TC4012.

Nábělek, J., Hetényi, G., Vergne, J., Sapkota, S., Kafle, B., Jiang, M., Su, H., Chen, J., Huang, B.S. (2009). Underplating in the Himalaya-Tibet collision zone revealed by the Hi-CLIMB experiment. *Science*, 325(5946), 1371–1374.

Panda, D., Kundu, B., Gahalaut, V.K., Bürgmann, R., Jha, B., Asaithambi, R., Yadav, R.K., Vissa, N.K., Bansal, A.K. (2018). Seasonal modulation of deep slow-slip and earthquakes on the Main Himalayan Thrust. *Nat. Comm.*, 9, 4140.

Pandey, M.R., Tandukar, R.P., Avouac, J.-P., Lavfi, J., Massot, J.P. (1995). Interseismic strain accumulation on the Himalayan crustal ramp (Nepal). *Geophys. Res. Lett.*, 22, 751–754.

Pandey, M.R., Tandukar, R.P., Avouac, J.-P., Vergne, J., Héritier, T. (1999). Seismotectonics of the Nepal Himalaya from a local seismic network. *J. Asian Earth Sci.*, 17, 703–712.

Pearson, O.N. and DeCelles, P.G. (2005). Structural geology and regional tectonic significance of the Ramgarh thrust, Himalayan fold-thrust belt of Nepal. *Tectonics*, 24, TC4008.

Perrier, F., Trique, M., Aupiais, J., Gautam, U., Shrestha, P. (1999). Electric potential variations associated with periodic spring discharge in western Nepal. *C. R. Acad. Sci.*, 328, 73–79.

Perrier, F., Chitrakar, G.R., Froidefond, T., Tiwari, D., Gautam, U., Kafle, B., Trique, M. (2002a). Estimating streaming potentials associated with geothermal circulation at the Main Central Thrust: An example from Tatopani-Kodari hot spring in central Nepal. *J. Nepal Geol. Soc.*, 26, 17–27.

Perrier, F., Gautam, U., Chitrakar, G.R., Shrestha, P., Kafle, B., Héritier, T., Aupiais, J. (2002b). Geological, geochemical and electrical constraints on the transient flow mechanism of the Dhor Barahi periodic spring in western Nepal. *J. Nepal Geol. Soc.*, 26, 109–119.

Perrier, F., Richon, P., Byrdina, S., France-Lanord, C., Rajaure, S., Koirala, B.P., Shrestha, P.L., Gautam, U.P., Tiwari, D.R., Revil, A. et al. (2009). A direct evidence for high carbon dioxide and radon-222 discharge in Central Nepal. *Earth Planet. Sci. Lett.*, 278, 198–207.

Perrier, F., Aupiais, J., Girault, F., Przylibski, T.A., Bouquerel, H. (2016). Optimized measurement of radium-226 concentration in liquid samples with radon-222 emanation. *J. Environ. Radioact.*, 157, 52–59.

Rai, S.M. and Upreti, B.N. (2006). Geology along the Arniko Highway between Barabise and Kodari (China-Nepal Boarder), Central Nepal. Unpublished, 1–10.

Richon, P., Perrier, F., Koirala, B.P., Girault, F., Bhattarai, M., Sapkota, S.N. (2011). Temporal signatures of advective versus diffusive radon transport at a geothermal zone in Central Nepal. *J. Environ. Radioact.*, 102, 88–102.

Robinson, D.M., DeCelles, P.G., Copeland, P. (2006). Tectonic evolution of the Himalayan thrust belt in western Nepal: Implications for channel flow models. *GSA Bulletin*, 118, 865–885.

Sharma, C.K. (1995). *Mineral Resources of Nepal*. Sangeeta Sharma, Kathmandu.

Shrestha, S.B., Shrestha, J.N., Sharma, S.R. (1984). Geological map of eastern Nepal, scale 1:250,000. Topograph. Surv. Branch., Surv. Dep., Kathmandu.

Stevens, V.L. and Avouac, J.-P. (2016). Millenary Mw>9.0 earthquakes required by geodetic strain in the Himalaya. *Geophys. Res. Lett.*, 43, 1118–1123.

Szeliga, W., Hough, S., Martin, S., Bilham, R. (2010). Intensity, magnitude, location, and attenuation in India for felt earthquakes since 1762. *B. Seismol. Soc. Am.*, 100, 570–584.

Tamang, S., Thapa, S., Paudyal, K.R., Girault, F., Perrier, F. (2019). Geology and mineral resources of Khudi-Bahundanda area of west-central Nepal along Marshyangdi Valley. *J. Nepal Geol. Soc.*, 58, 97–103.

Thapa, S., Tamang, S., Paudyal, K.R. (2018). Geological map of Khudi-Bahundanda area (1:25,000). Tribhuvan University, Kathmandu.

Thapa, S., Tamang, S., Paudyal, K.R., Girault, F., Perrier, F. (2019). Geology and micro-structure analysis of the MCT zone along Khudi-Bahundanda area of Lamjung District, west-central Nepal. *J. Nepal Geol. Soc.*, 58, 105–110.

Torres, M.A., West, A.J., Li, G. (2014). Sulphide oxidation and carbonate dissolution as a source of $CO_2$ over geological timescales. *Nature*, 507, 346–349.

Unsworth, M.J., Jones, A.G., Wei, W., Marquis, G., Gokarn, S.G., Spratt, J.E., the INDEPTH-MT Team (2005). Crustal rheology of the Himalaya and Southern Tibet inferred from magnetotelluric data. *Nature*, 438, 78–81.

Upreti, B.N. (1999). An overview of the stratigraphy and tectonics of the Nepal Himalaya. *J. Asian Earth Sci.*, 17, 577–606.

Upreti, B.N. and Yoshida, M. (2005). Geology and natural hazards along the Kaligandaki Valley, Nepal. Department of Geology Tri-Chandra Campus, Tribhuvan University, Kathmandu.

Wolff-Boenisch, D., Gabet, E.J., Burbank, D.W., Langner, H., Putkonen, J. (2009). Spatial variations in chemical weathering and $CO_2$ consumption in Nepalese High Himalayan catchments during the monsoon season. *Geochim. Cosmochim. Acta*, 73, 3148–3170.

# Conclusion

**Rodolphe CATTIN[1] and Jean-Luc EPARD[2]**

[1]*University of Montpellier, France*
[2]*University of Lausanne, Switzerland*

In this first volume, we have presented some elements to understand better the scientific approach used by researchers to unravel the many processes acting in the Himalayan dynamics. We have focused in particular on:

– the building of the Himalayan range and the Tibetan plateau;

– the large tectonic structures accommodating the India–Asia convergence;

– the geometry of lithospheric structures at depth;

– the slip rate along the Topographic Frontal Thrust in south Bhutan;

– the numerous hydrothermal systems of Nepal.

The second volume presents current studies describing the Himalayan range's main tectonic units. This information is key to estimating the dynamics of this region over the last 150 million years and assessing the conditions associated with the initial stage of the continental collision and the building of the structures presented in Volume 1.

# List of Authors

Lok Bijaya ADHIKARI
Department of Mines and Geology
Kathmandu
Nepal

Pierre AGRINIER
Paris Cité University
France

Nelly ASSAYAG
Paris Cité University
France

Théo BERTHET
Department of Earth Sciences
Uppsala University
Sweden

Mukunda BHATTARAI
Paris Cité University
France

Laurent BOLLINGER
French Alternative Energies and
Atomic Energy Commission
Bruyères-le-Châtel
France

Hélène BOUQUEREL
Paris Cité University
France

Fabio A. CAPITANIO
School of Earth, Atmosphere
and Environment
Monash University
Clayton
Australia

Rodolphe CATTIN
University of Montpellier
France

Carine CHADUTEAU
Paris Cité University
France

Marie-Luce CHEVALIER
Chinese Academy of
Geological Sciences
Beijing
China

Jamyang CHOPHEL
Earthquake and Geophysics Division
Department of Geology and Mines
Thimphu
Bhutan

Isabelle COUTAND
Dalhousie University
Halifax
Canada

Mathieu DELLINGER
Paris Cité University
France

Dowchu DRUKPA
Earthquake and Geophysics Division
Department of Geology and Mines
Thimphu
Bhutan

Guillaume DUPONT–NIVET
Géosciences Rennes
University of Rennes
France

Jean-Luc EPARD
University of Lausanne
Switzerland

Christian FRANCE-LANORD
University of Nancy
Vandoeuvre-lès-Nancy
France

Francesca FUNICIELLO
Department of Science
Roma Tre University
Rome
Italy

Vineet K. GAHALAUT
CSIR–National Geophysical
Research Institute
Hyderabad
India

Jérôme GAILLARDET
Paris Cité University
France

Ananta Prasad GAJUREL
Tribhuvan University
Kathmandu
Nepal

Eduardo GARZANTI
Università degli Studi di
Milano-Bicocca
Milan
Italy

Stéphanie GAUTIER
University of Montpellier
France

Frédéric GIRAULT
Paris Cité University
France

Vincent GODARD
CEREGE
Aix-en-Provence
France

Fanny GOUSSIN
University of Grenoble Alpes
France

Djordje GRUJIC
Dalhousie University
Halifax
Canada

Stéphane GUILLOT
University of Grenoble Alpes
France

Ratna Mani GUPTA
Department of Mines and Geology
Kathmandu
Nepal
and
Academia Sinica
Taipei
Taiwan

György HETÉNYI
Institute of Earth Sciences
University of Lausanne
Switzerland

Bharat Prasad KOIRALA
Department of Mines and Geology
Kathmandu
Nepal

Cécile LASSERRE
Claude Bernard University
Lyon
France

Rémi LOSNO
Paris Cité University
France

Kapil MAHARJAN
Department of Mines and Geology
Kathmandu
Nepal

Sudhan Singh MAHAT
Sanjen Jalavidhyut Company Limited
Kathmandu
Nepal

Gweltaz MAHÉO
Laboratoire de Géologie : Terre,
Planètes et Environnement
Claude Bernard University
Lyon
France

Konstantinos MICHAILOS
Institute of Earth Sciences
University of Lausanne
Switzerland

Kabi Raj PAUDYAL
Tribhuvan University
Kathmandu
Nepal

Frédéric PERRIER
Paris Cité University
France

François PREVOT
Paris Cité University
France

Anne REPLUMAZ
University of Grenoble Alpes
France

Thomas RIGAUDIER
University of Nancy
Vandoeuvre-lès-Nancy
France

Martin ROBYR
University of Lausanne
Switzeland

Taylor F. SCHILDGEN
University of Potsdam
and
GFZ German Research Centre
for Geoscience
Potsdam
Germany

Shiba SUBEDI
Institute of Earth Sciences
University of Lausanne
Switzerland
and
Seismology at School in Nepal
Pokhara
Nepal

Nabin Ghising TAMANG
Department of Mines and Geology
Kathmandu
Nepal

Shashi TAMANG
University of Turin
Italy
and
Paris Cité University
France
and
Tribhuvan University
Kathmandu
Nepal

Sandeep THAPA
Paris Cité University
France

Rasmus C. THIEDE
Christian-Albrecht University
Kiel
Germany

Bishal Nath UPRETI
Nepal Academy of Science
and Technology
Kathmandu
Nepal


Peter A. VAN DER BEEK
University of Potsdam
Germany

Jérôme VERGNE
University of Strasbourg
France


Shiguang WANG
Chinese Academy of
Geological Sciences
Beijing
China

# Index

# Summary of Volume 2

## Chapter 2. Suture Zone
Julia DE SIGOYER and Jean-Luc EPARD

## Chapter 3. Geological Evolution of the Tethys Himalaya
Chiara MONTOMOLI, Jean-Luc EPARD, Eduardo GARZANTI, Rodolfo CAROSI and Martin ROBYR

## Chapter 6. Oligo-Miocene Exhumation of the Metamorphic Core Zone of the Himalaya Across the Range
Rodolfo CAROSI, Salvatore IACCARINO, Chiara MONTOMOLI and Martin ROBYR

## Part 3. Lesser and Sub Himalayan Sequence

## Chapter 7. Lithostratigraphy, Petrography and Metamorphism of the Lesser Himalayan Sequence
Chiara GROPPO, Franco ROLFO, Shashi TAMANG and Pietro MOSCA

## Chapter 8. Sedimentary and Structural Evolution of the Himalayan Foreland Basin
Pascale HUYGHE, Jean-Louis MUGNIER, Suchana TARAL and Ananta Prasad GAJUREL

## Conclusion

Rodolphe CATTIN and Jean-Luc EPARD

# Summary of Volume 3

**Tributes**
Eduardo GARZANTI, Vincent GODARD, Rodolphe CATTIN, György HETÉNYI,
Jean-Luc EPARD and Martin ROBYR

**Foreword**
Rodolphe CATTIN and Jean-Luc EPARD

**Preface. From Research to Education: The Example of the Seismology
at School in Nepal Program**
György HETÉNYI and Shiba SUBEDI

**Part 1. Surface Process**

**Chapter 1. Orogenesis and Climate**
Frédéric FLUTEAU, Delphine TARDIF, Anta-Clarisse SARR, Guillaume LE HIR
and Yannick DONNADIEU

## Chapter 2. Eroding the Himalaya: Processes, Evolution, Implications
Vincent GODARD, Mikaël ATTAL, Saptarshi DEY, Maarten LUPKER and
Rasmus THIEDE

## Part 2. Natural Hazards

## Chapter 3. Glaciers and Glacier Lake Outburst Floods in the Himalaya
Christoff ANDERMANN, Santosh NEPAL, Patrick WAGNON, Georg VEH,
Sudan Bikash MAHARJAN, Mohd Farooq AZAM, Fanny BRUN and and
Wolfgang SCHWANGHART

## Chapter 4. Landsliding in the Himalaya: Causes and Consequences

Odin MARC, Kaushal GNYAWALI, Wolfgang SCHWANGHART
and Monique FORT

## Chapter 5. Himalayan Surface Rupturing Earthquakes

Laurent BOLLINGER, Matthieu FERRY, Romain LE ROUX-MALLOUF,
Jérôme VAN DER WOERD and Yann KLINGER

## Chapter 6. Seismic Coupling and Hazard Assessment of the Himalaya

Sylvain MICHEL, Victoria STEVENS, Luca DAL ZILIO and Romain JOLIVET

## Part 3. Focus

## Chapter 7. Recent and Present Deformation of the Western Himalaya

François JOUANNE, Jean-Louis MUGNIER, Riccardo VASSALLO,
Naveed MUNAWAR, Awais AHMED, Adnan Alam AWAN, Manzoor A. MALIK
and Ramperu JAYANGONDAPERUMAL

## Chapter 8. The 2015 April 25 Gorkha Earthquake

Laurent BOLLINGER, Lok Bijaya ADHIKARI, Jérôme VERGNE,
György HETÉNYI and Shiba SUBEDI

## Chapter 9. Crustal Fluids in the Nepal Himalaya and Sensitivity to the Earthquake Cycle

Frédéric GIRAULT, Christian FRANCE-LANORD, Lok Bijaya ADHIKARI, Bishal Nath UPRETI, Kabi Raj PAUDYAL, Ananta Prasad GAJUREL, Pierre AGRINIER, Rémi LOSNO, Chiara GROPPO, Franco ROLFO, Sandeep THAPA, Shashi TAMANG and Frédéric PERRIER

## Conclusion

Rodolphe CATTIN and Jean-Luc EPARD

Printed and bound by CPI Group (UK) Ltd, Croydon, CR0 4YY

17/08/2023

08101196-0001